"As ruins are to Rome, museums are to Paris, and theme parks are to Orlando, nature is to Monte Verde."

NPR radio program, *All Things Considered*

"We Costa Ricans have a tendency to see our attributes—democracy and peace—as gifts of God. But they're products of struggle. And biodiversity conservation must be the same."

Rector, University of Costa Rica

"We're in the business of biodiversity. It's like an art museum. It's the property of humanity. And there are certain pieces that you have to have."

Monte Verde Environmental Activist

"I'm already a conservationist. I don't need outsiders who come in with projects and money and screw things up. They destroy communities. I don't need their help. Call it a biological corridor, call it a reserve, call it a conservation area, call it what you want. It simply means more controls on the campesino."

Cattle Rancher, Monte Verde

# Contents

# Green Encounters

# Studies in Environmental Anthropology and Ethnobiology

General Editor: **Roy Ellen**, FBA
*Professor of Anthropology, University of Kent at Canterbury*

Interest in environmental anthropology has grown steadily in recent years, reflecting national and international concern about the environment and developing research priorities. This major new international series, which continues a series first published by Harwood and Routledge, is a vehicle for publishing up-to-date monographs and edited works on particular issues, themes, places or peoples which focus on the interrelationship between society, culture and environment. Relevant areas include human ecology, the perception and representation of the environment, ethno-ecological knowledge, the human dimension of biodiversity conservation and the ethnography of environmental problems. While the underlying ethos of the series will be anthropological, the approach is interdisciplinary.

# Green Encounters

Shaping and Contesting Environmentalism
in Rural Costa Rica

Luis A. Vivanco

*Berghahn Books*
New York • Oxford

First published in 2006 by
*Berghahn Books*
www.berghahnbooks.com

**Library of Congress Cataloging-in-Publication Data**

Vivanco, Luis Antonio, 1969–
  Green encounters: shaping and contesting environmentalism in rural Costa
Rica/Luis A. Vivanco.
    p. cm.—(Studies in environmental anthropology and ethnobiology; v. 3)
  Includes bibliographical references and index.
    1. Environmentalism—Costa Rica—Reserva del Bosque Nuboso de
Monteverde. I. Title. II. Series.

GE199.C8V58 2006
333.72097286—dc22                                           2006040746

**British Library Cataloguing in Publication Data**
A catalogue record for this book is available from the British Library


Printed in the United States on acid-free paper


ISBN 978-1-84545-168-4 hardback, 978-1-84545-504-0 paperback

# List of Figures

# List of Abbreviations

| | |
|---|---|
| ADISE | Asociación de Desarrollo Integral de Santa Elena (Integral Development Association of Santa Elena) |
| ANJEM | Asociación de Jovenes Ecológicos de Monte Verde (Monte Verde Association of Ecological Youths) |
| BEN | Bosque Eterno de los Niños (Children's Eternal Rain Forest) |
| CASEM | Comisión de Artesanos de Santa Elena-Monteverde (Santa Elena-Monteverde Artisans Commission) |
| CCT | Centro Científico Tropical (Tropical Science Center) |
| CTPASE | Colegio Técnico-Agropecuario de Santa Elena (Santa Elena Technical-Agricultural High School) |
| DGF | Dirección General Forestal (General Forestry Division) |
| GV | Global Volunteers |
| ICE | Instituto Costarricense de Electricidad (Costa Rican Institute of Electricity) |
| ICT | Instituto Costarricense de Turismo (Costa Rican Institute of Tourism) |
| ITCO | Instituto de Colonización y Tierras (Institute of Colonization and Lands) |
| MAG | Ministerio de Agricultura y Ganadería (Ministry of Agriculture and Livestock) |
| MCL | Monteverde Conservation League |
| MCFP | Monteverde Cloud Forest Preserve |
| MEP | Ministerio de Educación Pública (Ministry of Public Education) |

| | |
|---|---|
| MINAE | Ministerio de Ambiente y Energía (Ministry of Environment and Energy) |
| MIRENEM | Ministerio de Recursos Naturales, Energía, y Minas (Ministry of Natural Resources, Energy, and Mines) |
| MOPT | Ministerio de Obras Públicas y Transporte (Ministry of Public Works and Transport) |
| MVI | Monteverde Institute |
| OTS | Organization for Tropical Studies |
| RSE | Reserva Santa Elena (Santa Elena Reserve) |
| SINAC | Sistema Nacional de Areas de Conservación (National System of Conservation Areas) |
| SPN | Servicio de Parques Nacionales (National Park Service) |
| TSC | Tropical Science Center |
| UPANacional | Unión de Pequeños Agricultores Nacional (National Union of Small Agricultural Producers) |
| YCI | Youth Challenge International |

# Preface and Acknowledgements

On Tuesday 15 March 2005, five gunmen stormed, guns ablaze, into a bank in the rural community of Santa Elena, in the Monteverde region of Costa Rica. They intended to steal as much money as possible before escaping into the surrounding rain and cloud forests. It was not the first time this bank had been the target of a robbery attempt. During the past decade it had been struck two other times, and one of these assaults resulted in the death of one of the bank's security guards. What distinguished this third assault was the sheer bloodiness of it. Nine people died, among them four bank robbers, a policeman, and four bank clients. Several others were critically injured. One robber holed up in the bank with a hostage, a young bank employee, and a twenty-eight hour standoff with police ensued. The robber finally surrendered and released his hostage. The story made international news, and Costa Ricans, shocked, were glued to their television sets during the stand-off.

The tragic irony of this event is that it took place in a country renowned for a political culture of peace, and in a part of the country in which the religion of Quakerism has existed for over fifty years. It is a new kind of fame—or notoriety—for the Monte Verde region. The region is mainly known as one of the premier sites to study and experience high-altitude tropical cloud forest and its conservation in Costa Rica, and for some, the Americas (Forsyth 1988; Caufield 1991; Budowski 1992; Morrison 1994; Aylward et al. 1996; Baez 1996; Honey 1999; Burlingame 2000; Nadkarni and Wheelwright 2000). Within Costa Rica and in international environmentalist activism and scholarship, Monte Verde has long been a recognized leader in setting aside large areas of landscape for formal protection and as a pioneer ecotourism destination.

The formal protection of Monte Verde's natural patrimony goes back over thirty years, and is often identified as beginning with a small group of North American Quakers who settled there in 1951 to escape U.S. militarism, and the arrival of scientists to study the cloud forests (Caufield ibid.). Following the establishment of the Monteverde Cloud Forest Preserve on and near Quaker lands during the early 1970s, Monte Verde's visibility in environmentalist and scientific circles began to grow somewhat discontinuously until the late 1980s and early 1990s, when activists succeeded in placing it on the agenda of a transnational "Save the Rainforest" movement. At the height of these campaigns, literally hundreds of thousands of people participated efforts to formally protect Monte Verde's forests. Children in particular played a key role in these processes, and Monte Verde enjoys international fame as one of the first forests that children from many (at least fourty-four) countries helped "to save" (Patent 1996). Their contributions to organizations like the World Wildlife Fund, The Nature Conservancy, and an array of grassroots rain forest groups from the Americas, Europe, and Asia helped fund "adopt-an-acre" programs, forest guards, environmental education, and reforestation. In tribute to the children who helped protect the forests, one of them is named the Bosque Eterno de los Niños, or Children's Eternal Rainforest.

As a result, Monte Verde environmental organizations established Central America's largest private nature reserve (currently more than 22,000 hectares) and more than doubled the Monteverde Cloud Forest Reserve (to 10,500 hectares, although it is smaller now, as I will explain). Largely because of the international media attention this activity generated, Monte Verde has also become a sort of tropical mecca, attracting anywhere between (depending on which estimates you believe) 137,000 and 230,000 tourists and students per year—making it the second-most-visited area in Costa Rica—whose ecotourism is seen as a key strategy in support of biodiversity conservation and sustainable development (Aylward et al. ibid.; Burlingame ibid.; Baez and Valverde 1999; Ramirez 2004; ICT 2004).

The latest bank robbery is unlikely to cause a downturn in tourism, even though a tourist was among the injured. Even if they knew about it before visiting, most tourists would assume it is an unusual thing and not likely to be repeated. But the bank robbery raises uncomfortable questions about why an area like Monte Verde would be targeted three times during the past decade, each time with an increasingly violent outcome. Certainly, the bank has not been especially well-protected. When the first robbery took place in 1995, I happened to be conducting an interview nearby, and my friend and I walked over to see what was happening, not so wise given recent events. At the time, the bank had a young man as its security guard whose only weapon was a baton and whose main talent seemed to be talking to the pretty young women tourists who came to change money. There was only one police officer to serve the whole area. But the facts of our walking over to see what was going on, the loose bank security, and the area's minimal police protection say something about how distant the possibilities of serious violent confrontation appeared to Monte Verdeans. After this first robbery, an armed police officer was assigned to stand guard at the bank and the local police department was expanded and professionalized, although of course these facts did not stop the next two robbery attempts.

But bank protection and a professional police force are only part of the equation here. What is it about this particular area's level of economic activity that makes it attractive to bank robbers? And it is here that we see one of the little-discussed, maybe unexpected dilemmas of successful efforts to develop an economy based on nature conservation and ecotourism: the flow of large amounts of money into remote areas where there appears to be a good possibility of escape through wilderness areas. It is too easy and cynical to suggest that attracting the criminal element is a trade-off of this particular form of development, or maybe the accepted cost of "doing business" in ecotourism. Not only is this perspective downright disrespectful of the dead and injured, it also assumes that all Monte Verdeans have supported and participated equally in a regional development oriented toward formal landscape protection and tourism development. They have not.

In fact, as I explore in this book, Monte Verdeans have been debating the long-term environmental and social consequences of the pursuit of large-scale nature conservation and ecotourism for some time now. Struggles over conservation and ecotourism have taken the shape of public protests over the aims and methods of land purchases for formal protection; conflicts between landowners, tourism entrepreneurs, and environmental activists who disagree on techniques of land management; and even retribution against over zealous forest guards and conservation administrators. As one farmer once told me with considerable emotion, "I'm already a conservationist. I don't need outsiders who come in with projects and money and screw things up. They destroy communities. I don't need their help. Call it a biological corridor, call it a reserve, call it a conservation area, call it what you want. It simply means more controls on the *campesino*." More recently, even while expressing pride in their accomplishments, conservation and tourism leaders have publically expressed concern over the long-term environmental degradation and social divisions caused by uncontrolled ecotourism.

Over the years Monte Verde has received a lot of international media attention, and to a much lesser extent, scholarly attention in the social sciences. In most of this literature there is a tendency to emphasize the uniqueness and sheer triumph of Monte Verde's environmentalist achievements, and to downplay the social conflict that has been both an aspect of and generated by environmentalism itself. Many of these accounts emphasize that conservation goes against human history, and they often focus on the importance of Quaker ideologies of consensus and harmony in facilitating this change (Honey ibid.). Although I do not focus solely on conflict, this book's examination of points of tension and fragmented social processes, not to mention some highly polemical and locally controversial stories, is not meant to deny the existence of such ideologies or the fact that many people usually do their best to get along. But it does offer a provocative rethinking of certain events and processes based on voices and events typically not heard beyond Monte Verde itself, precisely because many of these previous accounts are either written as or sanctioned by elite or official environmentalist or tourism perspectives, with little attention to alternative perspectives. As a result, they tend to tell us the outcomes of environmentalism—how many hectares of forest are preserved, how many children have received environmental education, etc.—not how and why things happened the way they did, or whose voices were heard and whose were not heard, in the construction of environmentalist initiatives. It is from the vantage point of processes—not outcomes—that we can evaluate the role of environmentalism in ongoing changes in social institutions and peoples' lives, as well as the dynamic character of environmentalism itself.

The position I take is that it is useful to consider Monte Verde's experiences of environmentalism as encounters between people and institutions with disparate ideas and practices of nature and social change. As an arena of cultural encounter, environmentalism is a powerful new force of social and ecological change, yet its boundaries are fluid and those who claim environmentalist sensibilities are many. This approach entails a close description and analysis of conservation practitioners, as well as the diverse actors, institutions, and subjectivities with which they negotiate to implement their visions of social and natural change. The point is to interrogate how these diverse environmental philosophies interact and conflict with each other, in order to understand how environmental initiatives are actually shaped and contested in particular social and historical contexts. From this perspective, we have the opportunity to critically rethink understandings of environmentalism as a heroic struggle against the forces of history, and reframe it as an arena of social struggle and production in which certain ideologies of human-nature relations have gained prominence over others, and what the consequences of these processes are in a specific place like Monte Verde.

## Acknowledgements

Since I began going to Monte Verde over a decade ago, I have been fortunate to have encountered and been supported by some extraordinary people. First and foremost, I would like to express my profound gratitude to those people in Santa Elena, Cerro Plano, San Luis, Cañitas, La Cruz, and Monteverde in the Monte Verde region, and La Tigra in the San Carlos region, for their hospitality, willingness to share their lives, and in many cases friendship. I have often felt like an interloper and gadfly, especially among those who have been the subjects of research and international attention for the remarkable things they have said and done, and people have generally accepted me with grace, patience, and good humor.

Although I would like to thank Monte Verdeans here by name, I hope they will understand my decision to follow anthropological convention and maintain their anonymity. Moreover, if the previous reception of social scientific and anthropological research is any indication of the amount of controversy it generates, I find it even more necessary to protect the identities of those people who have been so helpful to me as I have conducted the research for the dissertation initially, and then this book. One of the difficult aspects of writing about Monte Verde is that certain stories in its environmental history are so well-known and widely published that it is a daunting task to assure the privacy of individuals involved in environmental activism there. I am sure that some would consider my unwillingness to expose individual identities as denying the Monte Verde region one of its most valued characteristics, which are the individual efforts and life stories of those who have pursued nature protection, which in some cases, have inspired people around the world. I appreciate this point of view, and even as I recount some of these efforts and stories, I know that other researchers and journalists will offer credit where it may be due. This confidentiality extends as well to the research assistants who I worked with in the mid 1990s, whom I hope I have thanked in private.

It is more difficult to conceal the identity of institutions. Not only does it seem impossible to hide the identity of institutions whose names circulate in global environmental circles, but I also believe that too many historical and sociological details would be lost if I changed their names. Therefore, I will take this opportunity to thank the board members, employees, volunteers, and members of the following institutions who offered me their time, perspectives, and logistical support in the course of this research, and many of which appear in this book: The Monteverde Institute (at which I was a research fellow in 1995-6), The Monteverde Conservation League, Reserva Bosque Nuboso de Santa Elena, the Colegio Técnico-Agropecuario de Santa Elena, Productores de Monteverde, Cooperativa Santa Elena, Monteverde Biological Station, Monteverde Cloud Forest Preserve, Tropical Science Center, Area de Conservación Arenal (since renamed Arenal-Tempisque), Asociación de Jovenes Ecológicos de Monte Verde (ANJEM), UPANacional office in Santa Elena, Asociación de Desarollo Integral de Santa Elena, Municipalidad de Monteverde, Monteverde 2020, and the office of Ministerio de Agricultura y Ganadería in Santa Elena. I was also fortunate to be received with collegial support at the University of Costa Rica's Instituto de Investigaciones Sociales and Anthropology Department.

I am grateful to the following institutions for their financial support. At the dissertation research and write-up stage I received support from: Wenner-Gren Foundation for Anthropological Research (Grant #5921), MacArthur Foundation, Mellon Foundation, Princeton University Department of Anthropology, Princeton University Council on International Studies, Princeton University Program in Latin American Studies, Princeton University Council on Regional Studies, New England Board of Higher Education, University of Vermont Graduate College, University of Vermont Anthropology Department, and the University of Vermont Program in Environmental Studies. As a faculty member at the University of Vermont, I was able to pursue further research and writing with the assistance of the University of Vermont's University Committee on Research and Scholarship. My most recent research in Monte Verde (2004) was supported by the Fulbright Scholars program, when I had the good fortune to serve as a teaching and research fellow at the University of Costa Rica's Master's Program in Social Anthropology.

The following colleagues have had a hand in reading drafts, offering advice, responding to my frustrations, or otherwise supporting this research project at its various stages, and to them I am grateful: Kay Warren, Rena Lederman, Emily Martin, Vincanne Adams, Ari Shapiro, Carol Zanca, Leslie Burlingame, Robert J. Gordon, Glenn McRae, Chuck DeBurlo, Keith Hollinshead, Joe Heyman, Felipe Montoya, Cindy Longwell, Jim Petersen, Richard Wilk, Martha Honey, Lynn Morgan, Brian Gilley, Bob Pepperman Taylor, Ben Minteer, Josh McLeod, Laura Cervantes, María Eugenia Bozzoli de Wille, and graduate students in the

University of Costa Rica's Master's Program in Social Anthropology. Special thanks to Marion Berghahn, Roy Ellen, Michael Dempsey, and Catherine Kirby for their support in the publication of this book.

My family has been a source of steadfast support throughout this whole process. My parents, Diane and Ed Vivanco, have been steadily interested in the unsteady progress of this book. But I am even more grateful to Peggy O'Neill-Vivanco, Isabel Vivanco, and Felipe Vivanco, who have been unfailingly supportive of this work and have given more meaning to this project than I could imagine. Because I broke the "no-book no-baby" policy twice over and she has suffered through this project with patience and humor, Peggy deserves the most respect and gratitude.

Small parts of this book have been previously published, and I gratefully acknowledge permission to reprint these parts from the publishers. Thanks to the University of Pittsburgh for permission to reprint parts of my article "Spectacular Quetzals, Ecotourism, and Environmental Futures in Monte Verde, Costa Rica," which appeared in *Ethnology: An International Journal of Cultural and Social Anthropology* vol. 40 no. 2, Spring 2001. Thanks also to Rowman and Littlefield Publishers, Inc. for permission to reprint parts of my chapter "Environmentalism, Democracy and the Cultural Politics of Nature in Monte Verde, Costa Rica," which appeared in *Democracy and the Claims of Nature: Critical Perspectives for a New Century*, R. Taylor and B. Minteer, eds. Rowman and Littlefield, 2002. Finally, thanks to Youth Challenge International for permission to reprint the poster that appears on page 140

# 1

## Introduction: Encounters in a Tropical Cloud Forest

Several weeks after he began a new job as a maintenance worker at a small cloud forest preserve in Monte Verde, Costa Rica, Manuel Azofeifa invited me to walk with him through the forest.[1] During our walk, Manuel explained that as a career cattle rancher and dairy farmer he had paid relatively little attention to the establishment and expansion of forest preserves and the growing numbers of foreign tourists visiting this remote mountainous region in which he and his family have lived and farmed for over fifty years. He decided to work at this preserve not so much because he wanted a midlife change of career or to profit in the tourism economy, but out of a desire to support his son's high school, which had acquired the cloud forest preserve in 1995 to fund its daily operations. We carried machetes and our lunch, typical on long forest walks like this, but Manuel viewed this walk differently from others he had taken, and not simply because he had not been in this forest since it was formally protected in the late 1970s, or that he now worked here as an employee. The shift in Manuel's approach to this walk was reflected in the camera he carried, a small point-and-shoot that he said he rarely uses except for special occasions like parties or family trips. This was the first time he had taken a camera with him into a forest.

When I asked him about it, Manuel explained that he brought his camera because he wanted to take a picture of a waterfall he knew was in a remote area to show his wife and daughters, since they would never come this far into the forest. When we found the waterfall he posed for me to take a picture of him in front of it, donning my Indiana Jones-style cowboy hat, for he said he wanted to look like he was on an *aventura* (adventure). Walking on, Manuel talked excitedly about waterfalls, saying that they are particularly what Costa Ricans enjoy when they visit *el bosque* (the forest). Then, on hearing the call of a resplendent quetzal, an iconic bird for Monte Verde tourists and environmental activists, he hurried ahead to see it. When I caught up to him, he was pointing his camera up a tree at the bird and making noises to get its attention. He snapped a photograph, turned to me, and declared, "You know, five years ago, I might have shot at this bird with a gun, probably just for the hell of it. But, you see, now I shoot it with my camera." (See figure 1.1)

*1.1 Seeing the forest with new eyes (photo by author)*

A year and a half later, as Manuel and I reviewed his family photo album, we landed upon the pictures of that day. He asked me if I remembered when we took these photographs. Although he had taken several others, only two photos were on display in the album: one, the image of Manuel posing in front of the waterfall with my hat, and the other a blurry image of dark green with no obvious bird. When I asked why he kept these photos in his album, particularly since one of them did not really seem to show anything (at least to my eyes, that were looking for a bird), Manuel told me that it was a way to remember "*la aventura que tuvimos aquel día*," or "the adventure we had that day."

## In the Heart of "The Green Republic"

Perhaps a camera seems a modest and commonplace object to carry on a walk through a tropical forest, especially in Monte Verde where since the late 1980s and 1990s the most visible forest creature often seems to be a camera-bearing tourist searching for a resplendent quetzal and other elusive cloud forest wildlife. But the fact that Manuel in particular, a man who does not fit any standard image of an ecotourist (usually upper-middle-class people from industrialized countries), framed and recorded his experience through a camera's lens implies a significant shift in how he relates to the forests he has long known and used.

In the worldwide battles to save distinctive natural landscapes and biodiversity, this otherwise unremarkable situation would nevertheless seem to mark a significant event. A Central American cattle rancher and hunter entered a cloud forest and, in the common phrase, engaged in a low-impact "take only pictures, leave only footprints" activity. Given that Manuel represents a key actor in the so-called "hamburger connection," the apparent chain of causation that begins with tropical forest destruction to produce inexpensive beef and ends with North American fast-food consumers, this event could inspire hope that the "end of nature" (McKibben 1989) is not a necessary outcome of modern development patterns and agriculture.[2] More importantly, in an era of (at least rhetorical) skepticism toward top-down development and environmental initiatives, it seems to demonstrate community-level acceptance of environmentalism and ecotourism.[3]

For those who are familiar with Costa Rica, a country dubbed "The Green Republic" (Evans 1999) for its commitment to sustainable economic development, nature conservation, and nature-oriented tourism, it may not come as much of a surprise to know that this happened in Monte Verde. Situated along the Continental Divide at around 1,200–1,400 meters altitude in the northwestern central Tilarán highlands, where moisture-laden winds from the Caribbean combine with Pacific winds to form a persistent cloud bank, the Monte Verde region is at the heart of the mythical Green Republic. It is the home of iconic tropical creatures, such as the resplendent quetzal and the endemic golden toad (*Bufo periglenes*), whose disppearance in 1989 is one of the world's great ecological mysteries (Pounds and Crump 1994). Aside from its high visibility as a place where North American Quakers settled in 1951 and as one of the most studied tropical cloud forests on record, it is one of Costa Rica's best-known sites of nature conservation, sustainable development, and ecotourism. One prominent Costa Rican scholar and environmentalist has declared, "The importance of Monteverde as a role model in the environmental movement of Costa Rica cannot be overstated. It is a living example of what could and should be done and what was done at the right time" (Gómez 2000: ix) (See figure 1.2).

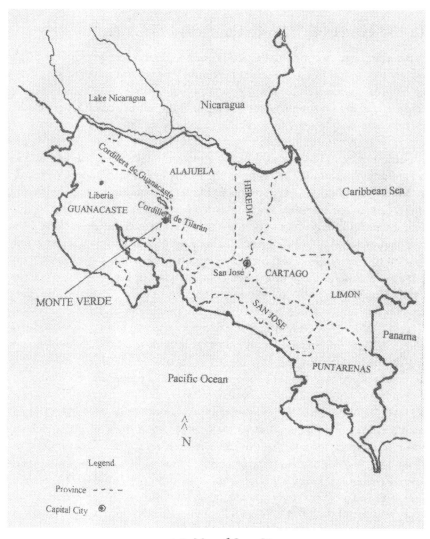

*1.2. Map of Costa Rica*

A central purpose of this book is to explore the key roles that environmental activism, biodiversity conservation, and ecotourism have played in ongoing changes in people's lives, identities, and futures in the Monte Verde region. One indication of these changes is the fact that the now-protected forests have been redefined as reservoirs of biodiversity and as spaces of touristic consumption, so when Manuel and others enter them, they do so under a new set of expectations and controls, backed by the power of national laws and local law enforcement, if not international expectations. Beyond the increased scrutiny on traditional land use practices, environmentalism has also contributed to changes in concepts of property ownership and forest management outside the formally protected areas. With the rise of nature-oriented tourism, it has fueled an economic transformation, rendering dairy farming (the region's main agricultural activity for the past four decades) of decreasing economic importance. This transformation has created new social, educational, and labor opportunities for residents and migrants attracted to this economy, including jobs as tourism guides, hotel work, and widespread environmental education for youths. There is also an ongoing realignment of political hierarchies and alliances, facilitating both a rise in "non-governmentality" (the key involvement of nongovernmental organizations in public affairs and regional governance) and Monte Verde's successful bid to gain municipality status in 2001. More broadly, environmentalism has provided the grounds upon which people articulate and create new social and political identities, reflecting changes in how many Monte Verdeans think about their relationships with neighbors, their place in their country, and their place in the wider world.

Common environmentalist and tourism narratives, as told to visitors and as written in the international environmental press, and even in natural and social scientific publications, tend to attribute the opportunities for conservation there to a unique social mix and grassroots identity (Caufield ibid.; Burlingame ibid.; Honey ibid.). The Quakers figure prominently in these stories, as romantic pioneers with the foresight and wisdom to establish a flourishing and peaceful community with nature protection as a central value. But so do scientists, who have begun to reveal the mysteries of the tropical forests and provide justifications for their protection (Collard 1997). These narratives privilege "Monteverde" ("Green Mountain"), the one-worded name that refers to the social and geographical space of the original Quaker village where the first formal conservation efforts were organized and directed. But they can be as uncritically celebratory of nature conservation's triumphs at the hands of enlightened Quakers and scientists as they are unidimensional and dismissive of the complex scenarios implied by the situation of my companion Manuel.

Manuel lives and works in the "other" Monte Verde, the two-worded name that encompasses the largely North American village of Monteverde, but also the several largely Costa Rican villages that surround it, as the municipality of Monteverde, Puntarenas province, with a current population of some four thousand people. Until it was declared a municipality just a few years ago and the one-worded name was officially adopted, the two-worded "Monte Verde" had been widely used by locals to distinguish the outlying areas from the Quaker village. It is a useful designation for this book because it entails certain sociogeographic distinctions under-stood locally. The political, economic, and social center of the wider Monte Verde is in Santa Elena (the village in which Manuel lives and works), but in practice, Monte Verde's definition can be even more expansive. It is defined by many as the communities in the "milk shed," a broader region with as many as fifteen or so vil-lages that spread across the mountainous highlands of two provinces (Puntarenas and Guanacaste) and fall within the economic sphere of the dairy factory in Mon-teverde village, and increasingly, the tourism economy.[4] Even though the one-worded Monteverde gets most of the journalistic and scholarly attention, it is the two-worded Monte Verde where most of the tourists stay in hotels, where many of the standing forests are and conservation efforts have been directed, and where most of those who work the land and hotels actually live. The destinies of "Mon-teverde," the historically Quaker and North American village, and "Monte Verde," the broader highlands region, are profoundly intertwined, and it is the latter that encompasses the former that is the main focus for this book.

Manuel's story is important precisely because of the ambiguity it indicates about environmentalism's meanings and achievements in Monte Verde. That day, he and I were both engaged in an adventurous encounter with a cloud forest. As attractive as it might be to view Manuel's behavior as a conversion to certain preconceived environmentalist values, it would be mistaken to assume that he and I share the same understandings of what we were looking at and doing, or even that Manuel consid-ers himself an environmentalist as you or I might think of it. Not only would this ignore the complexity and dynamism of Manuel's and my personal comprehensions of the landscape and its features, it would also assume that "becoming a *conserva-cionista*" is as instantaneous and uniform as putting on a shirt.[5] In fact, during our friendship, Manuel has continued to proudly work as a cattle rancher and proposed to me several times that we spend the night at the preserve where he worked to hunt for *tepezcuintle* (agouti), a large rodent prized for its tasty meat. A conscientious and knowledgeable collector of forest products, Manuel did not see any reason why he should not periodically, if discreetly, harvest wild hearts of palm and other forest products. He was aware that these behaviors offended certain environmentalist orthodoxies, if not were outright illegal, but he was unapologetic because he viewed these forests as familiar patrimony, not semisacred wilderness.

That Manuel also viewed this walk as an adventure is also significant, something that is not especially surprising given the privileged status of tropical forest wilder-nesses as an existential space of adventure in the modern imagination. However, if we take into account Simmel's famous observation that "… adventure, in its spe-cific nature and charms, is a form of experiencing. [But t]he content of the experi-ence does not make the adventure" (1971: 197), then perhaps Manuel's adventure reflects a practical shift in how he relates to this landscape: he was encountering the forest with, quite literally, new eyes, since he had a camera that he never had taken with him into a forest, shooting with it instead of a gun. By displaying his photographs, he may have been eager to demonstrate to himself, myself, and others who would see them that he could properly approach a cloud forest landscape and appreciate it as an ecotouristic attraction. But I soon discovered that Manuel did not necessarily agree with the designation of the forest as a protected area (he said he initially opposed it on the grounds, as some other farmers did, that it meant land being left out of agricultural production), or that the forest would always be maintained in this way in the future (he has been distrustful of tourism as a long-term economic solution), or even that he had a deep commitment to his job at the forest reserve (he quit after a year because of the poor salary). Manuel's actions that day represent new layers of meaning for him that are intertwined, even in tension, with his other attitudes toward cloud forests, the rural landscape, environmentalism, the tourism economy, and his identity as a *campesino*. From this angle, environmentalism is highlighted as powerful—it provides new boundaries and ways of framing human interactions with nature—yet it is also unfinished and dynamic, as divergent meanings are assigned to its practices and outcomes.

## Encountering Environmentalism

Recent social scientific depictions of Latin American environmentalism have tended to focus on the political and economic implications of ecological crises and nature protection campaigns for the nation-state (Fournier 1985; Bonilla 1988; Fundación Neotrópica 1988; Solera and Maldonado 1988; Bravo 1992; Fallas Baldi 1992; Segura 1992; Wallace 1992; Mora Castellanos 1993; Salazar et al. 1993; Collinson 1996; and Evans ibid.; Toledo 2000; Hilje et al. 2002; Roberts and Thanos 2003). Even as relatively uncritical celebrations of national environmentalist elites (i.e., Wallace ibid.; Hilje et al. ibid.), these studies have been useful for shedding light on the governmental and civil society arenas in which environmental policies are constructed. But most of these studies have tended to avoid the persistent and enormously interesting complexities of how environmental politics and cultures play out in settings like Monte Verde, where the contested landscapes and social practices that are destructive of nature actually exist. Furthermore, the voices and actions of rural Latin Americans are critically absent from these studies, especially the non-elite conservation workers and tourists, those socially situated actors whose everyday lives and interactions are key factors in how environmentalism actually works.

At a local scale, environmentalism is a complex social and institutional arena in which contradictory knowledges, meanings, and cultural practices play out (Milton 1996; Orlove and Brush 1996; Brosius 1997a; Thompson 1999; Selin 2003; Paulson and Gezon 2004). It is most certainly not a static or essential body of political and ecological issues. Environmental practitioners often make concerted efforts to address and incorporate perspectives and interests that they do not necessarily share, in order to gain converts and legitimacy, and so the boundaries of who counts and claims identity as an environmentalist are themselves dynamic (Brosius 1999). Environmental projects can be reworked and negotiated in terms that accommodate and help give shape to local identities and histories, as subjectivities and understandings of the environment themselves change. Furthermore, self-identified environmental activists are often at odds between themselves, struggling over the goals, scale, and meanings of their actions. As this book makes clear, these struggles are reflected in the complex expectations and relations between grassroots groups and their international allies and funders; shifting scientific knowledge claims and the uneasy incorporation of "local knowledge"; vigorous debates over how to define the "community" in projects that highlight community participation and benefit; intense disagreement over how to educate about environmental problems; and legal battles between local conservation groups fighting to gain control of the same land.

As "the environment" has become an important conceptual and practical category for Monte Verdeans, it has extended people's understandings of their worlds, created new subjectivities, and redefined social institutions and relationships (Agarwal 2005). A central goal of this book is to show how these processes take place in spaces of cultural encounter, that is, contingent sites of cultural interchange and projection. The notion of encounters suggests spaces of cultural interaction, conflict, and production in which different views of the world come into consciousness, visibility, and dialogue—are shaped and contested—though not necessarily (if ever) on equal footing (Adams 1996). To frame this study as an exploration of encounters means that this is not an ethnography of "a people" in the traditional anthropological sense. Or, for that matter, of conservation practitioners solely. If we view environmentalism as an arena of encounters between diverse actors, knowledges, and interests, each based on pluralistic constructions of what is nature in the tropics and how to manage it, the ethnographic focus is upon the encounters themselves as specific communicative interactions. In my use here, encounters are not objective constructions, but relational, dialogical, performative, and experiential spaces that exist in the context of everyday interactions, social relationships, and institutions. Methodologically, the focus on encounters encourages a descriptive emphasis on processes of becoming and dynamism—of shaping and contesting, of collaboration and marginalization—more than focusing on essentialized things and categories.

It is significant that the encounters featured in this book reflect specific world-historical conditions in which the boundaries between the "local" and "global" are also in flux. Environmentalism itself mediates between these categories, projecting globalizing aspirations and abstractions while being practiced in concrete places (Ingold 1993). An important ethnographic task is to characterize the "jagged articulations" (Clifford 2001) between places, people, and moving products, and therefore the ethnographic encounters in this book are not purely or simply "local" but entail actors and discourses that move in and out of the geographic and cultural spaces of Monte Verde. This requires, as both Ingold (ibid.) and Tsing (2000) have cautioned, moving beyond a framework (often seen in the anthropology of globalization) in which globalization and its mechanisms (in particular "circulation") represent transhistorical processes that rest on a dichotomy of dynamic "global forces" and stable (if less and less isolated) "local places" (Tsing ibid.; Piot 1999). Even as such a framework draws valuable attention to the political and economic processes that drive social interconnection and inequality in the contemporary world (projecting Hannerz's famous "global ecumene"), it often downplays the collaborations, misunderstandings, oppositions, and dialogues that take place in environmentalist projects, not to mention the specific ways in which actual people make meaning of their places in broader political, economic, and cultural systems.

In the end, this argument for seeing environmentalism as an arena of relational encounter is meant to focus attention on how the category "environment" has become an increasingly key site of cultural production in peoples' everyday lives, where certain privileged ideas about natural and social realities are channeled, giving shape to new relationships between and among natures, nations, individuals, and institutions (Brosius 1999). Encounters is a different kind of metaphor than "Fortress Conservation" (Brockington 2002), which has recently gained some currency in political ecology and the cultural study of environmentalism. There is a similar thread here in the sense that these two approaches take for granted that environmentalism is an apparatus of power in which there are clearly winners and losers, provoking social conflict and exacerbating existing inequalities. But for Brockington and others pursuing the fortress metaphor and its variations, especially although not solely in the context of East Africa (Bonner 1993; Neumann 1998; Guha 2000; Anderson and Berglund 2002; Igoe 2004), the emphasis is on describing how local experiences with nature conservation are the product of European colonial ideas and institutions that connect the civilizing mission and a landscape devoid of humans. The results include systematic native dispossession and exclusion, as well as intractable conflicts between indigenous peoples, wildlife, and natural resource managers.

There are clearly elements of subordination and domination in Monte Verde's encounters with environmentalism, including the desire to expel (certain) people from the landscape, a combination of state policies and international funding that have helped achieve those ends, and the generation of intense conflict over nature and its uses. But the inadequacy of the fortress metaphor is that it projects a ready-made edifice of impenetrable walls plunked down in an alien context. Emily Martin's acute observation about science studies' tendency to view science as a "citadel on a hill" is relevant here, and we need just substitute the word "scientist" with "conservationist" to appreciate her point: "These studies share something in common with the perspectives of scientists themselves about the primacy and power of the knowledge that they are producing. Scientists ... are always the active agents in these scenarios: they translate, read, write, mobilize, impose, convince ... they are always pictured as active agents attempting to change an essentially passive world" (Martin 1994: 6). One of the lessons of Monte Verde is that environmentalism is not a monolithic force intervening in a passive world, but an internally diverse set of ideas and practices that operate in a complex social environment whose boundaries are not so fixed as to be impenetrable. A central goal of this book is to illustrate some of the specific instances and processes in which those ideas and practices are shaped and contested.

## "No Artificial Ingredients": Knowing and Saving Tropical Nature

Monte Verde's privileged status as a conservation and tourism icon within Costa Rica itself can also tell us a lot about how one particular nation has experienced a particular form of globalization in its rural areas: "green development" (Adams 1990). Costa Rica is a global showcase for a certain environmental narrative: it is a biodiversity hot spot (5 percent of the world's floral and faunal species reside there, and it is thought to have the highest species diversity density in the world; Obando 2002: 14) that is threatened with destruction by local people; but the tide is being turned and tropical nature redeemed by progressive individuals and organizations promoting its formal conservation (Campbell 2002), not to mention a tourism economy in which almost 67 percent of its 1.2 million annual visitors go to protected areas (ICT 2004). In fact, along with other touted political and cultural attributes, especially stable liberal democracy and peace ("the Switzerland of Central America"), Costa Ricans have recently incorporated ecological appreciation as a positive national characteristic. Indeed, as one writer has quipped, "When Costa Ricans want to sell something, they paint it green" (Rovinski 1991: 56).

By virtually any standard, the scale of conservation activity and investment in Costa Rica is impressive. From the 1970s to the present, Costa Ricans have vigorously pursued the creation and expansion of national parks, formally protected wilderness areas, and a state environmental bureaucracy. Currently, 25.1 percent of the national territory is under some category of formal public protection, with an additional estimated 5 percent under private protection regimes, the largest percentage of national territory under legal protection in Central America (Herlihy 1992; Rosales 1997; Obando 2002: 47).[6] Tourism based on the attraction of protected rain forests surpassed bananas and coffee as the primary source of foreign exchange in 1995, and the national legislature has passed progressive laws governing forestry, wildlife conservation, biodiversity, and a constitutional reform guaranteeing all citizens "the right to a healthy and ecologically balanced environment" (Salazar 1993).

The most extraordinary aspect of this story, some observers note, is that the country has undertaken its conservation efforts in a context of limited public financial resources due to debt crisis and three rounds of IMF structural adjustment during the 1980s and 1990s (Sun 1988; Evans ibid.). A crucial aspect of this success was the country's visibility as an island of peace and democracy in a war-torn region, and President Oscar Arias's Nobel Peace Prize for his regional peace initiatives during the 1980s is commonly credited with putting Costa Rica on the map and laying the groundwork for the tourism economy (Rolbein 1989; Honey 1994a). But also crucial is Costa Rica's visibility as an environmental problem in industrialized countries where there has been a growing awareness of tropical deforestation, and the fact that transnational environmental and civic organizations have at one point or another invested substantial financial resources in support of Costa Rican conservation efforts and bureaucracies.[7]

Ironically, at the same time that it gained its reputation as an international leader in conservation and nature-based tourism, Costa Rica experienced one of the highest rates of deforestation in Central America.[8] The general consensus is that since the 1950s, the rate has fluctuated between thirty thousand to sixty thousand hectares per year, or 2.5 to 3.9 percent of the territory per year (Hartshorn et al. 1982; Utting 1993; García 2002).[9] Called by one environmental historian "The Grand Contradiction" (Evans ibid.), this situation reflects the ongoing structural complexities of agrarian social change, neoliberal development policies that favor ecologically destructive nontraditional agricultural exports, inconsistent government enforcement of environmental laws, timber extraction, and the very politics of environmentalism itself (Thrupp 1981; Anger 1989; Thrupp 1990; Vandermeer and Perfecto ibid.; Edelman ibid.; Viales 2002). Highlighting the conflictive nature of these processes, the Rector of the University of Costa Rica said at a 1997 conference, "We Costa Ricans have a tendency to see our attributes—democracy and peace—as gifts of God. But they are the products of struggle. And biodiversity conservation must be the same."

Although it is a marketing scheme, the Costa Rican Institute of Tourism's (ICT) motto "No Artificial Ingredients" offers an apt metaphor for the country's image as a conservation success story, and why complex situations like Manuel's have typically been marginalized, if not seen as irrelevant, in discussions about tourism and conservation there. Even though the ICT's advertisements always describe Costa Rica's people as an idealized "friendly," they are simply that, with little or no elaboration. It is the country's endowment of tropical nature *without people* that is important, and the real reason to visit Costa Rica is for its natural splendor and the adventure of visiting tropical wilderness. But it is not simply tourism that has marketed tropical nature and underrecognized complex Costa Rican social contexts. In telling their own histories, environmentalists have also tended to oversimplify the cultural and political milieus in which they practice their politics, confirming Hobsbawm's famous observation that the least appropriate historian of railways is a rail enthusiast. For example, two of Costa Rica's most prominent environmental activists observe, "Conservation is congruent with the political and social values of the country.... It can be argued that Costa Rica, with its natural resources and traditional values supported by international academic and conservation organizations, has established a model system for the preservation of tropical diversity" (Gámez and Ugalde ibid.: 134). It also often follows that Costa Rica's identity as the Green Republic is the result of an educated and literate populace that understands the practical importance of ecology, and the "stability of an unarmed democracy and its satisfactory attention to the basic socioeconomic needs" of its people (Fournier 1985: 33; Gámez 1991). The emphasis is on tropical nature's transcendence over a simplified social history, as if there is something *essential* about Costa Rican nature and *inevitable* about its conservation.

Consider a U.S. National Public Radio series that examined the current state of environmentalism and ecotourism in Costa Rica (Burnett 1997). Its purpose was to investigate how, in spite of its international reputation as a model for nature conservation, Costa Rican environmentalism was confronting internal challenges, including continued deforestation, apparent government indifference, and failed ecotourism projects. One segment examined Monte Verde specifically, in which correspondent John Burnett focused on a small dairy farming community in the region called Monte de los Olivos that has been struggling to attract visitors to its cooperatively managed hotel and small forest reserve. Even while highlighting Monte Verde's reputation for successes in conservation and ecotourism (he says conservation is "contagious" there), the report offered a frank appraisal of how the promises of sustainable development had not borne equal fruit for all, and described the dilemmas faced by residents who had committed themselves to the tourism economy with poor results.

But Burnett's description of Monte Verde drew from deeper currents. He said, "As ruins are to Rome, museums are to Paris, and theme parks are to Orlando, nature is to Monte Verde" (ibid.). Burnett's comparison places Monte Verde in a category that includes some of the world's great tourism destinations and icons of civilization, but with one important difference: the first three represent monuments of human artistic expression, accomplishment, and entertainment, while Monte Verde is distinguished for that which is apparently not human, or is independent of human creation. But understanding the shape of environmentalism in Monte Verde is not as simple as asserting that it reflects the inherent beauty or biological importance of Monte Verde's flora, fauna, and ecosystems. Such arguments are common among environmental activists, tourism operators, and tourists who have promoted Monte Verde as a pristine natural destination (Gómez ibid.). There are several problems with such explanations, one being that they do not adequately address why other places with similar attributes—cloud forests with resplendent quetzals, for example, which exist throughout Costa Rica and Central America—have not also been designated in international media and activism circles as great sites of nature. They also do not direct us to understand the cultural processes through which Monte Verde has become a showcase for a specific kind of nature, *tropical* nature, as a hyper-nature whose destruction and conservation themselves become hyper-visible.

Therefore, we must situate Costa Rica and Monte Verde's joint rise in global conservation circles in relation to broadly circulating conversations and ideas about nature in the tropics, its destruction, and its revaluation as something worth saving. Although it is enjoying less prestige as a global political and civic problem than it was in the 1980s and early 1990s, "saving the rain forest" continues to be widely viewed in the North as a virtuous and progressive act, for the benefit of the world if not also the people who live in the landscapes targeted for conservation. In a broad sense, the transnational political space that opened around saving the rain forest—in which Monte Verde occupies an important place—has tried to address the ecological destruction wrought by capitalism and agricultural expansion in the South, and consumption-oriented lifestyles in the North. Coinciding and usually intersecting with "sustainable development," one of its central goals has been to create an alternative form of economic development in the tropics with ecological stewardship and wilderness preservation as core values (Head and Heinzman 1990; Caufield ibid.). As Escobar (1997) suggestively observes, the problem of rain forests is also a consequence of what he calls the "irruption of the biological." That is, because the very survival of biological life itself is at stake, capital, science, and politics have responded in ways that now define nature as something "worth saving." Nature in the tropics is attractive not only because it symbolizes biological profusion itself, but because of its increasing value as a reservoir of tourism, biodiversity, and biotechnology prospecting and research.

To be sure, in recent decades, there has been a proliferation of literary, visual, and scientific tales of nature in the tropics, and their narratives tend to emphasize both its astonishing richness (using the imagery of "green gold") and rapid, tragic disappearance. Since the 1960s and 1970s, investigative journalists, activists, and scientists have published studies and "reports from the field," many for popular audiences, that describe the destruction and its causes, sanctifying the role of witness to a global crisis (Myers ibid.; Caufield ibid.; Head and Heinzman ibid.; Place 1993; O'Connor 1997). Often framed by descriptions of the mystery and exuberance of rain forests, these narratives emphasize the identification of culprits (Ellis 1996)—including population growth, ignorance, greed, poverty, consumerism, and the like—and the normalizing assumption that their disappearance is a problem for everyone, invoking (among others) health (as a reservoir of potentially miraculous cures), food security, soil erosion, and the air we breathe (as carbon sinks and the lungs of the world). The depressing imagery of destruction has become especially common in video documentaries and other mass media representations of rain forest ecosystems and creatures. Indeed, the images of destruction—it seems they always feature a poor and dirty peasant forcing a chain saw or bulldozer on majestic and vine-interlaced trees, then setting them on fire to clear the land for agricultural production—often serve to visually explain their futures (Vivanco 2002a).

The destruction of tropical forests *is* dramatic and troubling in its scale, as the now well-known statistics make clear. Scientists estimate that tropical rain forests cover up to 8 percent of the world's land surface (roughly thirty million km2), and between half and 90 percent of the total number of the world's approximately one and a half million species reside or are found exclusively in them (Park 1992: 6). By 1990, perhaps as much as half of the world's tropical forests had already been lost, with an estimated 20 percent of that loss occurring just in the thirty years between 1960 and 1990 (Terborgh 1992: 187). Between 1950 and the 1980s, Costa Rica experienced the disappearance of as much as 50 percent of its forest cover because of agricultural expansion and logging (Hartshorn et al. 1982). Tropical forests are also key sites in the so-called "Sixth Great Extinction": it is probable that a tropical rain forest species goes extinct every half-hour, an annual total of 8,760 species, and that 41 percent have already become extinct (Park 1992: 84; Morell 1999). Efforts to document and quantify the loss of natural resources, environmental services (air quality, water retention, and so on), and effects on global climate change suggest equally dramatic consequences (Carranza et al. 1996).

Notwithstanding the inclination to be alarmed, I think it has to be said that when they remain disembodied, many of these narratives and statistics of destruction appear *exaggerated*. By identifying their claims as exaggerated, I do not mean to suggest that they are empirically false or to discredit as fraudulent. Rather, I am assuming the word's root meanings, "to carry" or "to heap." Taking these and other claims about tropical forests seriously means recognizing that they carry—are heaped with—cultural, intellectual, and moral assumptions about how to think about and measure nature and its destruction, assumptions that need to be disentangled and placed in the relevant historical and social milieu in which certain kinds of claims about global realities are accepted over others. For example, tropical forests exist all over the globe, and often embedded in these narratives is the reductionistic tendency to use one specific place or story to represent the universal phenomenon of rain forest destruction. For a time, as I will discuss later, the story of Monte Verde's cloud and rain forest landscapes became iconic of *all* tropical destruction, overlooking or downplaying the always complex local causes and consequences of deforestation (Rudel and Horowitz 1993). Similarly, in a climate of crisis and urgency, narratives of degradation often implicitly carry standardized solutions that provide the basis of political action. The result can be the simplification of highly complex, multideterminate situations to fit a predetermined message and short-term solution erroneously believed to be universally applicable (Roe 1991; Leach and Mearns 1996; Campbell ibid.).

As Nancy Stepan explores in her recent work *Picturing Tropical Nature* (2001), beyond the sheer rapidity of its destruction, there are deep cultural currents that help explain the intense contemporary visibility of tropical nature. Borrowing the term "tropicality" to emphasize its discursive character, Stepan shows how in the last several hundred years artists, writers, scientists, journalists, travelers, and others in the West have imaginatively constructed tropical nature as an autonomous space with its own characteristic plants, animals, people, and diseases. With the rise of natural and social field sciences and colonial medicine in the nineteenth century, tropical nature became a zone of imagination, adventure, and inquiry in relation to which the attributes and meanings of (among others) civilization, human evolution, modern rationality, and medicine have been defined.

Stepan usefully characterizes the often contradictory symbolic meanings that have been projected onto nature in the tropics and the uses to which they have been put, in order to understand how nature has come to represent the tropics itself, that is, nature *as* the tropics and the tropics *as* nature. But she also correctly points out that tropical nature is a heterogeneous thing, "a mix of the natural and the artificial, the human and the non-human, the organic and the non-organic, both a physical space of living and non-living things and a human invention" (ibid.: 242). That is, its most basic quality is its dual nature: on the one hand, it serves to describe an empirical place with physical realities and order; on the other, it provides a projective arena through which social groups and individuals mobilize and (re)define categories relevant to the social orders in which they are situated. As an autonomous zone of purity in which human relationships with it are often perceived to be either insignificant or overwhelming, tropical nature has come to exist as a synecdoche for nature (landscapes, flora, fauna, ecosystems, biodiversity, and so on) in all of its forms. Furthermore, because of its status as a symbol of biological profusion and fragility, tropical nature has lately gained privileged status as *the* visual and intellectual barometer of nature's loss and conservation around the world.

Such imaginings have important consequences for how the problem of "saving the rain forest" has been understood and justified in places like Monte Verde. Undoubtedly, tropical conservation is prestigious work, perhaps even an *axis mundi* of global environmentalism. As one of the recent popularizers of rain forests, Norman Myers, writes (1984/92), tropical forests represent "the primary source of all life" and the last great location and laboratory of the world's biological diversity. In this vision, biologically strong tropical nature is the antithesis of the biologically weak temperate zones, a mythical and actual space where the scientist's dream to discover a new species and to revel in complex ecosystems can be realized. It is hyper-nature, or, simply, Nature itself, writ large. And so conservation, as a result, is also writ large: what more crucial place to save nature than in its most intensely rich, but also intensely threatened, site? This attitude generates a sense of enfranchisement: as one Monte Verde biologist-conservation activist told me, "It's like an art museum. It's the property of humanity. And there are certain pieces you have to have."

Yet there is no unified environmentalist vision or discourse of what that tropical nature is and should be, much less what to do with it. It is, as I explore in this book, assigned a wide variety of meanings and uses, some overlapping, others conflicting: wilderness, habitat, watershed, a space of recreation, a space of adventure, tourism spectacle, an exhibit hall for environmental education, laboratory, refuge of spiritual inspiration, a woodlot, a factory of nontimber forest products, to name but a few. Furthermore, at the center of tropical nature politics are ideological and iconic differentiations between peoples (Hollinshead 1999). This is often perceived as the modern and progressive rationalities of (often Northern) environmentalists butting up against the primitive and backwards rationalities of (often Southern) rural people. And there has been an increasing political awareness that in the realm of sustainable development, "justice" and "nature" often collide, and when they do, Northern elites and their allies promoting nature often dominate over Southern majorities desiring justice (Sachs 1996). To some extent, this is because the proximity of rural and forest peoples to forests makes them easy targets to blame for the destruction, and they often lack the political or social capital to refute these charges (Dove 1993). Furthermore, well-funded and politically connected environmentalists insistent on saving nature by taking lands out of production through the exclusionary apparatus of protected areas and reserves can represent yet another obstacle to fair distribution of lands and agrarian reform (Vandermeer and Perfecto 1995; Guha 2000).

These insights raise important questions about the cultural and political aspects of saving the rain forest, all of which motivate this book. What does it actually mean to "save the rain forest?" What actions are considered effective? Who is saving the rain forest, and from whom and what is it being saved? What is the source of environmentalist representations of tropical peoples and natures? Who identifies problems and develops solutions? On what intellectual and moral authority are their claims based? For whom do they apply? The language of "stakeholder inclusion" is commonly applied to the problem of rain forest conservation, the accepted actors including scientists (often differentiated according to their specialties), natural resource administrators, entrepreneurs, government bureaucrats, recreation managers, local nongovernmental organizations, international development and environmental organizations, and the broad category of "local people." But the normalizing language of stakeholders, itself a product of universalistic and bureaucratic rationality that has depoliticizing effects (if not intentions) (Brosius 1999), provides an inadequate lens to understand the variety of cultural perspectives and the conflicts that emerge over how tropical nature and its conservation are defined in practice, because it is based on a simplification that identifies actors according to their competing interests alone.

## Overview of the Book

This book explores different aspects and consequences of the emergent "culture of nature" in Monte Verde. It is narratively organized around specific encounters that illustrate distinct dimensions of and struggles over environmentalism. It is based on participant observation and archival research in Monte Verde environmental institutions, public and private spaces, protected areas, and tourist sites over the course of a decade, from 1993 until 2004, with the longest period in 1995–7. One advantage of this long-term involvement is that this study does not take a slice of environmentalism at a particular moment and make conclusions based on that slice. If it did, this book would not be able to show the enormous shift in the status of certain environmental institutions in Monte Verde since I began research there in the early 1990s—when international funding was prevalent—to today, when funds are scarce for conservation and ecotourism dominates the scene. But while I am interested in showing the social, economic, and political contexts in which the environment has emerged as a problem in Monte Verde, this book is not simply about the "artificial ingredients" side of the "No Artificial Ingredients" equation. Distinctions of this kind obscure the complex mix of human and nonhuman elements that underlie how and why tropical nature has become the object of intense scrutiny and its role in redefining Monte Verde's social orders and futures.

The book's focus on encounters has certain important features. First I write in a self-consciously processual and antiessentializing language, a deliberate strategy meant to complement the theoretical and organizational focus on dynamic and contested social fields. I do not use this language to suggest that all is in flux and constantly fractured, as some critiques (and caricatures) of postmodern writing imply. On the contrary, I will show that even while environmentalism may be a contested social field, specific groups and factions do succeed in imposing their vision of how to relate to the natural world through specific projects and programs.

Secondly, this book is not an exhaustive catalogue of every environmental initiative that has existed or the debates they have inspired, focusing instead on emblematic encounters in which people and institutions with multiple ideological agendas, knowledges, and methods negotiate the meanings and practices of environmentalism. My examples are selective, yet not arbitrary. One of my main concerns here is to track and characterize key interactions and processes in which powerful orders—especially science, tourism, development, and rural communities—have been increasingly reshaped and contested through the apparatus of environmentalism. These encounters can be divided into roughly three distinct arenas: the formal protection of landscape, outreach efforts to engender popular participation, and the pursuit of tourism as an environmentally sensitive economic development strategy. So the book is divided into these three sections, each with its own introduction, which even while they stand independently of each other, share common elements and themes. Each introductory section provides a theoretical framework that unites the disparate topics treated in the chapters.

The first section of the book examines how environmentalism became focused in Monte Verde on a certain version of "saving nature" by formally establishing preserves. It asks how tropical nature became a high visibility problem in Monte Verde, how Monte Verde itself became a high visibility problem for people around the world concerned with the loss of rain forest landscapes and biodiversity, and the struggles surrounding the establishment of formally preserved areas. This entails bringing both the natural-historical and social-historical conditions of life in Monte Verde into the same frame, in order to understand how they are intertwined. Beyond a discussion of Monte Verde's agrarian history (and the evolving diversity of perspectives on appropriate agriculture there), I explore how activists have understood the problem of environmental degradation and its solutions, and what intellectual, institutional, and material resources they have drawn upon to support their efforts to redress this degradation. It is here where I discuss the astonishingly successful, but also highly controversial, concentration of lands as formally protected nature preserves.

In the second section of the book, I explore the emergence of what I call environmentalism's "social work," which is intertwined with formal landscape protection in the era of sustainable development. The goal of this section is to understand how environmental institutions and activists have sought to directly remake the social and economic behaviors and subjectivities they identify as threatening to tropical nature, as well as to consider the significance of sustainable development initiatives for the people whose lives have been targeted as problems requiring such intervention. There is implicit, if not practical, importance in hearing and trying to understand the voices and experiences of the nonelite people whose participation is sought by these programs. It is no small matter that in our conversations, Manuel expressed mixed feelings about environmentalist representations of rural peasants as destroyers of nature. On the one hand, he told me, sometimes with embarrassment and sometimes with pride, detailed stories of changes in the landscape over his own lifetime, the product of clear-cutting and the expansion of dairy production throughout Monte Verde. On the other, he expressed subtle understandings of landscapes and ecosystems that many scientists would reject as undisciplined ignorance or superstition, but which obviously had empirical validity for his ability to relate to the landscape without irreversibly destroying it. Although he often spoke admiringly of Monte Verde's reputation as a model for conservation and ecotourism, Manuel had also participated in several protests against environmental groups working in the Monte Verde region, partly because he felt this was something being done *to* him and his community, not *with* them, even with elaborate "social work" programs under way. So a key goal of this section is both to assess how sustainable development has understood and intervened in the lives of people like Manuel, and to document the oppositional politics and alternatives that have been proposed both within and at the margins of mainstream environmentalism.

In the third section, I examine the opportunities and dilemmas of ecotourism in Monte Verde, which has emerged as a powerful framework for how residents and visitors alike have increasingly begun to view and interact with regional landscapes. Monte Verde is one of the most prestigious sites in the global ecotourism industry, and as such the region has experienced both the power and dystopias of ecotourism to reshape peoples' lives and landscapes. The resplendant quetzal, and the various efforts to present it for tourist consumption, symbolizes the ongoing shift in the productivity of Monte Verde's landscapes. As Manuel himself expressed, it is a shift that requires a new way of looking—quite literally looking—at landscapes now resignified as spaces of visual appreciation, consumption, and adventure. Moreover, ecotourism marks the application of new regimes of productivity to nature that require attention to a logic of spectacle and performativity, through practices literally inscribed on landscapes (through trail construction, the creation of viewing platforms, and so on). At the same time, tourism has expanded so rapidly and spun off new styles of interacting with landscapes (including canopy tours where visitors zip through trees), that Monte Verdeans are grappling with the ambiguous social and ecological effects of their tourism economy, and some are beginning to openly discuss the irony that their tourism economy itself threatens to undermine the very reputation Monte Verde has for "living in harmony" with tropical nature.

In the conclusion, I reflect on what has resulted from these encounters with environmentalism, which Monte Verdeans and environmental advocates beyond Monte Verde will be dealing with and trying to make sense of for years to come. Indeed, Monte Verde is currently at a crossroads, as it grapples with its new municipality status and its latest tourism boom. Because there is no singular environmentalism, but multiple institutions and activists with diverse intentions and practices that have changed their priorities and practices over time, this is not a simple matter of declaring that a certain number of hectares have been "saved," that a certain number of farmers planted a certain number of trees in reforestation programs, or that a certain number of children have sat through environmental education presentations. It is much more about considering, from diverse perspectives, the often confounding benefits and dilemmas that green encounters have provoked and amplified in the ongoing sociocultural and ecological changes taking place in the region, and what they might illustrate about the broader phenomenon of environmentalism itself.

# Notes

1. Manuel Azofeifa, like all the personal names used in this book, is a pseudonym to protect the privacy of the individual.
2. The "Hamburger Connection" thesis is a well-established orthodoxy in Central American environmental politics. See Nations and Komer (1987). For a more critical reading, see Edelman (1992).
3. One representative perspective states, "Conservationists, economists, and tourists have awakened to the realization that you can't save nature at the expense of local people. As custodians of the land, and those most likely to lose from conservation, locals should be given a fair share. Sound politics and economics argue for making local people partners and beneficiaries in conservation, as opposed to implacable enemies of it" (Lindberg and Hawkins 1993: 8).
4. This includes the communities of Monteverde, Cerro Plano, Santa Elena, La Lindora, La Guaria, Guacimal, San Luis, and Los Llanos in Puntarenas province; and Cañitas, La Cruz, Las Nubes, San Rafael, Los Tornos, Cebadilla, Campos de Oro, San Bosco, Cabeceras, San Ramón del Dos, La Florida, La Chiripa, El Dos Arriba, El Dos Abajo, Monte de los Olivos, Río Negro, Turín, and La Esperanza in Guanacaste province (Griffith et al. 2000: 390).
5. A note on terminology is necessary here. Locally, Costa Ricans refer to environmentalism as *la conservación*, or conservation. They do not use the term *el ambientalismo*, which is a more accurate translation of environmentalism. In English, "conservation" tends to refer to practices of landscape protection, while "environmentalism" is a broader social movement seeking to redefine human relations with the natural world (Gottlieb 1993), of which conservation is but a subset. Local meanings of *conservacionista* include both the specific sense of landscape protection and the broader social movement. I use the term *conservacionista* here, to reflect that this is what my companion would call it.
6. The government-protected categories, each with its own legal status and characteristics, include national parks, biological reserves, protected zones, forest reserves, wildlife refuges, indigenous reserves, and biosphere reserve (Herlihy 1992: 36–7). Currently, 12.21 percent of the national territory lies in 15 national parks, the strongest category of protection (Obando ibid.: 48).
7. For example, between 1961 and 1990, WWF sponsored eighty-two projects in Costa Rica, representing an investment of over $4 million, based on what it viewed as "a key opportunity to extend wildlands in Central America" (WWF 1992) (but also knowing that it would receive extensive coverage in the environmental media in North America and Europe; WWF 1986; WWF 1988). Between 1991 and 1999, 50 percent of the country's protected areas were established, largely with international donations (Obando 2002: 46). Beginning in the 1990s, Costa Rican biodiversity also attracted multinational pharmaceutical companies like Merck and British Pharmaceutical, which have signed agreements with the Instituto Nacional de Biodiversidad (INBio, or National Biodiversity Institute) to collect, preserve, and process samples of the country's plants and insects for their commercial potential as medicines, cosmetics, and so on. Although critics have challenged these relationships as experimental schemes undermining national sovereignty in favor of foreign-based extractive industries with few concrete benefits for Costa Ricans (Rodriguez 1992; Marozzi 2002), they have received extensive international media coverage as positive examples of sustainable development (Janzen 1991; Aylward, et. al. 1993; Gámez 1999).
8. There has been a lively debate over the quantities of forest lost, a debate that has intensified as technologies for measuring forest cover from satellite imagery improve (Escofet 1998).
9. For example, one relatively recent study, using satellite (LANDSAT) imagery to measure increases in forest fragmentation, proposes that between 1986 and 1991, the average rate was a remarkably high 4.2 percent (Sanchez-Azofeifa et al. 1997).

# Part One: Monte Verde's Visibility

"Any determinate and hence demarcated space necessarily embraces some things and excludes others; what it rejects may be relegated to nostalgia or it may be simply forbidden. Such a space asserts, negates, and denies."

(Lefebvre 1991: 99)

When I became friends with him during the mid 1990s, José Castellanos had just begun a joint venture with a friend of his to grow vegetables on a small plot of land on the outskirts of Santa Elena. Their plan was to sell the vegetables to Monte Verde's ecolodges and hotels, whose number had mushroomed over the past decade. José no longer owned any land where he could undertake such a venture, so the land belonged to his friend. The reason he did not own land, José often told me regretfully, is that he had sold it to an environmental group based in Monteverde, which was aggressively buying land in the region during the late 1980s and early 1990s. Until 1992, he owned seventy-five hectares of rain forest wilderness with his brother about six or seven kilometers north of Santa Elena, in an area called San Gerardo Arriba. Even now, but especially when he was a boy growing up there, San Gerardo is known as a remote area of difficult access and even more difficult possibilities for agricultural production, due to high precipitation and lack of a decent road. He never had any illusions about the difficulty of living there, but he had great passion for the landscape and its intricacies. José was known to some as a *baqueano,* a rural Costa Rican term for someone who is highly knowledgeable about forests, often because they are hunters.

On our walks to and from his friend's farm, José delighted in telling me stories about the plants, animals, and birds we saw in the forest patches along the road (see figure 2.1). But since he sold his land, he had not returned to San Gerardo and did not have any interest in visiting the forest reserve that had been formally established there. His encounter with environmentalism left a bitter taste in his mouth. He was profoundly resentful that he had to sell his land, telling me that he was forced into doing so because his brother with whom he owned the land was in debt and needed cash. He was angry at his brother's profligate ways, but even more at the conditions and institutions that he saw as ultimately responsible: the designation of an extensive area, including San Gerardo, as a national hydroelectric reserve in 1977, and the environmental groups who opportunistically exploited the landowners' willingness to sell their lands at low prices. He was especially frustrated by the fact that, after generations of official incitements and legal inducements to colonize the agricultural frontier, his land was legally frozen: he could not longer cut trees, and the banks would no longer loan money using land as collateral.

When environmental activists showed up with substantial funds to buy land, José and other landowners felt they had no choice but to sell. This sell-off led to the virtual disappearance of San Gerardo as a community, a theme I will explore in more detail in Chapter Five. Even though he barely understood the earnest concern over his patch of rain forest in the Northern countries from which most of this money flowed, José clearly perceived that he was no longer welcome there once he sold it. He told me, "There will always be national parks and protected areas, but we the people who lived there before are no longer welcome there. Conservation closed the forests, so I don't go there anymore. Plenty of tourists see it because they have the money. I don't have any contact with the tourists, but I see that they come with a lot of money. Many people are making business on these forests right now. Where does that leave me?"

*2.1. José on his way to the fields (photo by author)*

José's story offers a compelling place to open the first part of this book, because land purchases are perhaps Monte Verde's most consequential encounter with environmentalism. It highlights both the transformative influence of land purchases for formal protection in Monte Verde's economy, society, and landscape, as well as the perception among people like José that they did not necessarily choose the conditions under which they sold their lands. Although these are inseparable aspects of tropical nature's emergence as a high-visibility problem in Monte Verde, they do not necessarily tell us how environmental groups gained control over large areas of landscape, processes that were facilitated by important transformations in rural Costa Rica related to structural adjustment, the failures of social democracy at the national level, and increasing transnationalization and urbanization of rural economies (Edelman 1999; Viales ibid.). It is the purpose of Part One to consider these social, natural, and political-economic conditions, and to present how a certain normative version of nature, unpeopled tropical nature, has become intensely visible in Monte Verde. What is at stake here is not simply who lives in and interacts with tropical nature, but who has the authority to define those human-nature relationships in the first place, over the claims of others who also interact with those landscapes.

As Lefebvre observes in the quotation that opens this introduction, the redefinition of landscape from a working space endowed with social history to unpeopled nature preserve is predicated on assertive acts of inclusion, negation, and denial. In the new demarcation of space implied by the concentration of lands as nature preserves, certain actions, histories, and people—especially people like José—are forbidden or relegated to nostalgia, or as Lefebvre suggestively observes, rendered "obscene." Using the example of the façade, Lefebvre elaborates, "A façade admits certain acts to the realm of the visible, whether they occur on the façade itself (on balconies, window ledges, etc.) or are to be seen *from* the façade (processions in the street, for example). Many other acts, by contrast, it condemns to obscenity: these occur behind the façade" (Lefebvre ibid.: 99). The idea that landscape can be separated from human history, to (re)constitute nature's independence of human interference is the visible façade, both a condition for and an outcome of the creation of nature reserves. Because this is a historical act, it is important to understand what and who is rendered "obscene" by the new normative façade and how this is accomplished, and to understand the traces of the past (and present) that remain inscribed there.

An analysis of Monte Verde's tropical natural façade and the obscenities it has rendered can contribute insights into environmentalism's ideological and political status in the ongoing transformations that have been taking place in rural Latin America during recent decades. Anthropologist Michael Kearney (1996) has argued that social scientific representations have accorded peasants and *campesinos* a special status in discourses on the Latin American countryside. These representations have tended to circulate as reified images of peasants as static semiprimitives, justifying their inclusion in (and domination through) preconceived development agendas. But, as Kearney rightly contends, the people often identified as peasants do not fit the image of the static and autonomous small-scale farmer. Reflecting ongoing transformations in Latin American nation-states due to neoliberal reforms and transnational migration, Kearney argues that "post-peasants" are just as concerned with their linkages to land as they are globalized politics of human rights, ethnicity, migration, structural reforms, and ecology (ibid.: 8). It certainly was the case for my companion José, who had migrated around Costa Rica several times in his life, participated in land invasions on large farms with other peasants, and developed an articulate critique of ecological crusaders.

In certain ways the changes Kearney has in mind are present in Monte Verde's changing agrarian economy, as globalized rain forest politics and tourism have redefined its rural communities as "buffer zones" of nature preserves. Such politics have thrust people like José onto an international stage in debates over the future of tropical nature, forcing them to renegotiate their status in the countrysides in which they live, as well as their identities as *campesinos*. But as Kearney counsels, "The point here is that students who wish to critically assess the position of *the* peasants in rural society must first come to terms with *the images* of them." (ibid.: 59). Sustainable development and environmentalist discourses of rurality and projections of local peoples provide a privileged vantage point to do this, for they tend to highlight rural peoples as needy and semideveloped, whose destructiveness and lack of development require intervention in order to save nature.

In fact, the authority of environmentalist claims for control and stewardship depends upon the idea that the inhabitants of a place do not properly relate its natural resources, and that more appropriate forms of human-nature interaction exist. Even though environmental discourse has permitted the authority of indigenous knowledge and practices through green primitivism and "the ecological noble savage"—the presumed ecological wisdom of indigenous peoples (Ellen 1986; Vivanco 2003)—*campesinos* of European descent have typically been perceived as lacking the cultural or racial distinction that provides them with the authority and knowledge to sustainably manage tropical nature, and often defined in terms of their ignorance and as destroyers of nature. Some Costa Rican residents of Monte Verde admit as much, such as a biologist's assistant, himself born on a Monte Verde dairy farm, who observed: "Conservation is just not in our culture, so we have destroyed nature." This has also led to an interesting, if regularly unacknowledged, contradiction: while natural resource management perspectives increasingly stress that local people represent essential *resources for* the proper management of tropical nature, those people are also often characterized as *obstacles to* its proper management because of their destructive ways and ecological ignorance.

A major purpose of Part One, then, is to explore the gap between images of *campesinos* and the actual lives people have led, a gap that I believe has laid the groundwork for the symbolic legitimation of agendas for absolute nature protection, as well as tensions over the roles of *campesinos* as participants in those processes. In the first chapter, I offer perspectives on settlement patterns and agrarian histories of the region, offering a kind of political-ecological alternative to mainstream accounts of rural Monte Verde history that have projected (using Lefebvre) an "obscene" of "destructive *campesinos*." Under these symbolic projections lie important diversities in land management styles, as well as structural changes in Costa Rica's agrarian situation that facilitated environmentalism's rise in Monte Verde.

The second chapter looks at the emergence of environmentalist practice directly, focusing on processes of internationally funded land purchases and preserve establishment. They are not the only techniques environmentalists have employed to pursue their goals of sustainability, but they are arguably the most controversial and consequential in helping environmental organizations to achieve an unprecedented concentration of land and to legitimate their status as important new public authorities and political actors. Environmentalists have tended to see preserve creation as mainly a technical act. But establishing forest preserves is a profoundly moral and political act, with important consequences for the conceptual and practical relevance of biodiversity conservation and ecotourism in Monte Verde, precisely because of what it renders visible, invisible, and obscene.

# 2 Monte Verde's Agricultural Environment

"Everyone sees forests differently. Some see them as biologists, others as medical doctors. But I see them as a *campesino,* because that's what I am. And so I see that people need to survive with these forests. We can't not use them. We must use them wisely, though."

Dairy farmer, Santa Elena

In their discussions on biodiversity loss and landscape degradation, environmentalists tend to point to *campesinos* like José as responsible for a large part of the destruction. They explain that poverty, ignorance, greed, fast population growth, even apparent cultural predisposition are the root causes that drive people like José to rapidly cut rain forests. For example, in her popular book on the global situation of rain forests, *In the Rainforest,* Caufield has a chapter on the Quakers of Monteverde ("Cattle in the Clouds") in which she complains that the unchecked ambition to destroy rain forests there is the product of "the Iberians' overwhelming atavistic urge to become a caballero" (1984/91: 121), or cattle rancher. So we could add irrational inflexibility and static attachment to tradition to the list of possible root causes.

Perhaps because "saving the rain forest" is widely considered to be a modern and progressive activity, rain forest activists tend not to think of themselves as encumbered by mythologies, which in their modernist meaning suggests superstition and ignorance. But ideas and stereotypes about environmental degradation do not necessarily serve as neutral reflections of natural realities, much less their social, political, and economic causes and consequences (Blaikie and Brookfield 1987). Especially in contexts of perceived crisis, they can serve to mobilize powerful negative stereotypes and objectifications of rural peoples and provide justifications for standardized blueprint policy actions (Roe 1991). In fact, certain myths (understood in their anthropological sense, not as falsehoods, but as important foundational narratives and images that help define a meaningful basis for action) play an important role in perceptions of the problems and solutions for rain forest disappearance. One such powerful myth is that environmental problems are even reducible to "root causes" or singular variables, and the search for primary causes and the assignation of blame have represented central axes of debate in polemics over the destruction of the world's tropical rain forests (Vandermeer and Perfecto ibid.; Ellis ibid.).

In Costa Rica, the expansion of the agricultural frontier based on deforestation by small-scale farmers, and the consequences of these processes for biodiversity, have become isomorphic with environmental degradation itself (Radulovich 1988; Thrupp 1990). This selective lens has often been at the expense of mainstream attention to other ecologically and socially destructive activities like pesticide abuses resulting from inadequate regulation of multinational agricultural and chemical corporations, as in the case of the banana or coffee industries (Hilje et al. 1987; Saez 2003). Given the apparent immediacy of small-scale farmers to deforestation, it has also laid the grounds for objectifying rural peoples as destructive, as the "obscene" against which the scene of rain forest's redemption is defined. As I will show in both this and the next chapter, the mythology of the "destructive *campesino*" provided both a compelling explanation for deforestation and all the problems associated with it as well as a key justification for the removal of those people from the landscape through land purchases. But as an explanation for deforestation, it obscures crucial heterogeneity in *campesino* involvements with and attitudes toward the landscape, as well as important structural and political-economic transformations that influence the shape of those landscape interactions.

In Monte Verde, small-scale and subsistence farmers have been key participants in deforestation, although most of the deforestation ended over thirty years ago. Knowing they participated, though, does not by itself help us understand the factors that contribute to causes of deforestation, and their often important comparative differences from neighboring provinces and regions. Important factors that contribute to deforestation include local ecology (difficulty of access, levels of precipitation, diseases), economic factors and limitations like access to markets, political and legal factors like land tenure laws and conventions of property ownership, and cultural attitudes about the landscape and its uses. Furthermore, as Marcus Colchester reminds us, "It is both wealth and poverty which underlie deforestation and it is in the inequitable structures which link the two that the roots of forest loss can be located" (1993: 4). We should therefore consider how local causes of deforestation reflect and amplify external political-economic spheres, such as national agricultural policies, international development aid agendas, and hemispheric foreign policy relationships that favor certain forms of natural resource intensive economic development.

# Settling the Green Mountains

Sixty years ago, "Monte Verde" did not exist. That is, the Quakers did not yet live there, and so it did not have that name. But the area they named "Monteverde" (Green Mountain), in honor of the verdant highland tract they purchased from a mining company in 1951, did already sustain a sparse and decentralized population of Costa Ricans who had been settling these highlands since at least the 1920s, if not before. This last point is something not made in most environmentalist and ecotourism representations of Monte Verde history, in which history almost invariably begins in 1951 with the Quaker arrival (Burlingame ibid.; Griffith et al. 2000; Rachowiecki 2002). The Costa Rican settlers thought of themselves as residents of a *caserío* (neighborhood) of Guacimal, a once-active gold mining community about fifteen kilometers down the mountain (most tourists still pass through the sleepy village on their way to and from Monte Verde, oblivious to its previous importance as a highland commercial center). The people originally called where they lived Espinero, and later Santa Elena and Cerro Plano (Flat Hill) (see figure 2.2).

Aside from its fertile soils for subsistence agriculture, sugar cane, and small-scale coffee production, the area was renowned as a haven for producers of contraband liquor, who sold sugarcane *guaro* to the gold miners below. The emergence of the two-worded "Monte Verde" as a regional identity goes hand in hand with the arrival of the Quakers and the important influence they have had on life in the region since the 1950s, from the dairy economy they established to the first steps they took toward formal forest conservation with scientists and environmental activists.

Recounting stories about the Monte Verde region's pre–Quaker history is a partial and tentative affair, because there are few written or reliable records that might indicate who owned land, where property boundaries lay, or even who lived there. Not only did individual landowners often resist registering land claims and transactions with the National Registry to avoid paying taxes, there has also often existed a gap between land titles and who actually lived and worked the land because of informal rental schemes, boundary conflicts, land invasions, and outright intimidation. Indeed, even with various efforts to systematize land ownership records, it was not until environmental groups began purchasing and registering lands that many ownership records have been clarified and registered in the region.[1] The picture I present here is based on oral accounts from *moradores* (pioneers), most of whom have now passed away, and their middle-aged children, supplemented with various documentary sources and academic studies. As has been shown elsewhere in Latin America, patterns of settlement, forest conversion, and farm preparation were constituted by myriad factors: population densities, commodifiable or useful natural resources, ecological factors like disease and precipitation, national policies and legal traditions, integration into regional markets, and attitudes toward forests and agriculture (Barlett 1982; Fernandez 1989; Hecht and Cockburn 1990; Jones 1990;

Place 1993; Porras and Villareal 1993; Rudel and Horowitz 1993; Painter and Durham 1995; Bedoya and Klein 1996; Nygren 1995).

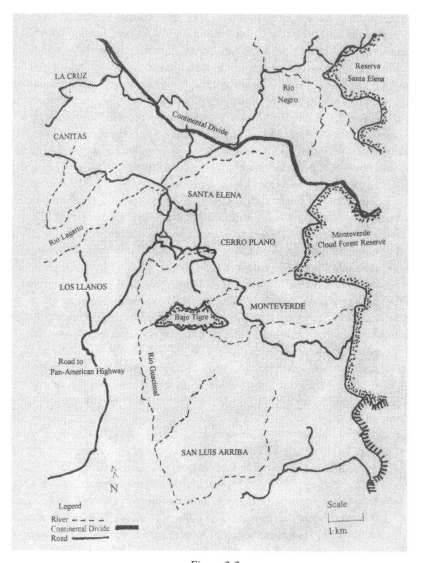

*Figure 2.2*

The initial movement of *campesinos* into the Tilarán highlands has its basis in socio-economic processes in the northwestern central part of the Costa Rican Meseta Central (Central Valley), and the expansion of mining activity in Guanacaste during the last decades of the nineteenth century. According to Samper

(1985), during this period the mercantile economies of the western Meseta Central cities of San Ramón, Naranjo, and Palmares in Alajuela province (from which many contemporary Monte Verde Costa Ricans trace their ancestry) experienced a gradual fragmentation as appropriate lands for production became exhausted and the technical and social abilities of family farms to specialize production and absorb extra labor decreased. Further undermining family farms was the consolidation of monopolistic control over the processing and commercialization of coffee production by the political and economic elites often identified as the "Coffee Oligarchy" (ibid.: 81–2; Escobar 1977; Gudmuson 1983; Castro and Willink 1989). By the late 1880s, sustained migrations of *Cartagos* (people identified as having geographic and cultural origins in the Central Valley) were establishing settlements in the forested Tilarán mountains of Guanacaste province, especially around Tilarán, approximately forty kilometers from where contemporary Monte Verde lies (Edelman 1992: 146; Hilje 1987). Turn-of-the-century migrants to the highlands tended not to be pauperized individuals, but people who sought recently cleared lands to reproduce production patterns in which they had participated before migrating, including sugarcane, coffee, and beef cattle production (Hilje 1987: 147; Sewastynowicz 1986).

By 1910, the lands around Tilarán had already begun a process of concentration so that fewer people controlled larger and larger holdings. Reportedly, tenure conflicts increased, and the migration of *Cartagos* from the western Meseta Central brought surplus labor to the gold fields of Abangares. The clearing of forest in the highlands was one way those people found work (Hilje ibid.). Throughout the 1920s and 1930s, *Cartago* families also entered the region, enticed by economic incentives from the municipality of Cañas to settle its territory as well as the renowned fertility of highlands soils, including some who began pushing their way toward the area of Santa Elena to establish coffee farms.[2] Others began entering the region from the direction of Puntarenas and Esparza to the south, settling in the San Luis valley just below the future village of Monteverde. By 1930, an Austrian "Scientific Commission" reported that the Tilarán highlands were already largely deforested as a result of this colonization: "The forests have been stripped of their grandest trees, because here, as in all parts of the country, cutting is done without foresight, measurement, criteria or control" (Hilje ibid.: 15).

Another factor attracting migrants was mining. In his study of Costa Rican agrarian history, Sandner (1964) acknowledges the significant connection between gold mining and colonization processes in highland Guanacaste. Although gold mining has never been an extensive economic activity in the Monte Verde region itself, some early settlers were *coligalleros* (prospectors), and virtually all settlers came from and interacted with the mining economies of Las Juntas and Guacimal.[3] In 1901, in fact, the areas of Santa Elena and Cerro Plano fell within a five-thousand-hectare land claim granted to the international mining interest Guanacaste Syndicate, Ltd., although the uplands near the Continental Divide were not extensively mined (Hilje ibid.).

In Monte Verde, settlement processes were nondirected, in the sense that they did not follow any rubric set by authorities as in formal colonization programs, as in the latter half of the twentieth century throughout the country (Seligson 1980). To *moradores* (pioneers), the forested and remote highlands represented *tierras baldías,* the category of uncultivated public lands whose meaning implies "common" or even "wasted" lands. Their claims to possession-by-occupation were solidly based on Costa Rican law and tradition. As a way to colonize its territory and a politically expedient alternative to land reform in the more densely populated Meseta Central, the Costa Rican state and municipalities provided incentives and land tenure laws to encourage the colonization of the agricultural frontier (Augelli 1987). For example, in 1852, colonists could claim up to 450 hectares. By 1939, with the passage of the Ley de Tierras Baldías, that number had been reduced to thirty hectares (Hall 1985). Under a 1942 "Ley de Precaristas" ("Squatters Law"), squatters who possessed privately owned land for more than a year were given possessionary rights (Hartshorn et al. 1982: 28). Claims to land possession rested upon the ability to show that it was being *mejorado* (improved), or put to productive use by being cleared and cultivated. In fact, until the 1996 passage of a new national forestry law, the value of land in rural Costa Rica was established by the areas cleared of forest and cultivated, not the amount of forested land (up to a 50 percent difference), so settlers sought to convert at least some of their forests relatively rapidly to establish their claims (Nygren 1995).

In the area around Santa Elena, the period of relatively open *cogiendo de derechos* (grabbing rights, as settling lands in rural Costa Rica is often called) did not last beyond the 1940s. According to one woman, whose family arrived in Cerro Plano during the 1920s,

> Seventy years ago, all you had to do was get the signature of the person who owned the land to occupy it. Nowadays they call that squatting, but these people were the pioneers of this area, so they took what was available to them. Even when I was a young woman [in the 1950s] the President of the Republic encouraged us to take lands and make them productive. By the time the Quakers arrived in the early 1950s, even though this was still a very isolated place, people had already begun to at least get their lands measured and boundaries were pretty well established. There were no fences, but people have more or less always known what was theirs. There were always fights over land boundaries, and a lot of violence. A bill of sale was your right to possession of land, so when you bought land this is what established your ownership. It still is. Very few people actually ever had their lands inscribed in the national registry.

Unlike North American concepts of property ownership based on an absolutist and static assumption of domain and control, in rural Costa Rica access to land was conceived as "rights," and land not in production or cultivation could belong to whomever claimed it (Seligson ibid.). It is therefore appropriate to view notions of private property in the Monte Verde region in terms of labor investment, but also their flexibility and negotiability, trends that would eventually clash with absolutist property concepts assumed by environmental groups.

The arrival of the Quakers in 1951 marks the beginning of an important regional transformation. The story of the Quakers is by now a central part of the Monte Verde mythos, and so is quite well-known (Caufield ibid.; Howard 1989; Mendenhall 1995; Honey 1999; Burlingame ibid.; Album Committee 2001). After spending months searching for lands to purchase in rural Costa Rica— Costa Rica was desirable to the Quakers who were seeking to escape U.S. militarism, since four of the young men in the group had just been jailed for resisting the draft—the group eventually decided to buy rights to a 1,400-hectare tract of semi-deforested uplands between Cerro Plano and the Continental Divide. They paid $50,000 to the Guacimal Mining Company, then owners of the land claim, and paid money to Costa Rican settlers (the Quakers still call them "squatters") who already lived there. The lands were divided depending on the needs of each family, but the settlers agreed to set aside a 554-hectare plot of steep cloud forest as a watershed for the Guacimal River. Though it was not exactly a commons (it was controlled by the family that helped raise the money to buy the land), community members nevertheless recognized that this forest was important as a woodlot and watershed for hydroelectric energy. During their first years there, the Quakers hired Costa Rican laborers to help cut forest that had not yet been cleared for pastures, paying them money, as opposed to barter, which had been a prevalent mode of exchange. They also introduced automobiles (helping to create the road to get them in and out), chain saws, a pelton wheel, medicines, and high-quality dairy cattle, with the intention of establishing a dairy industry. They relied heavily on the knowledge of Costa Rican settlers about vegetables and pastures that grew in the area, the pests that existed and how to control them, the identification of useful woods for fuel and construction, hunting advice, how to manage oxen, and how to speak Spanish, among other things. Among resident Costa Ricans, there continues to be great respect for the Quaker settlers, who were seen as hardworking people who brought new technologies and economic possibilities for residents without trying to convert the predominantly Catholic residents to the poorly understood religion of Quakerism.

Although none of them were professional dairy farmers before Monteverde, the Quakers considered cheese production an attractive economic base, because cheese could withstand the difficult roads to market, and ecological conditions in the highlands favored dairy cattle over beef cattle (Griffith et al. ibid.: 393). In 1953, eight of the Quakers formed a stockholding dairy corporation and cheese factory, Productores de Monteverde. Within several years, it was apparent that the Quakers could not produce sufficient milk on their farms, so Costa Rican farmers were invited to contribute milk for processing.[4] By 1961, Costa Rican producers outside of the Quaker settlement were already contributing the same amount of milk as the Quakers themselves, although the actual number of Costa Rican milk producers was higher since each produced smaller quantities than Quakers. Beginning in this period, milk and cheese production became the central economic activity in the region, and for the first time many Costa Rican farmers were able to access a regular source of income. Today there are over two hundred producers spread across multiple highlands communities who provide milk on contract to the factory. The factory produces at least eight cheeses, ice cream, and sour cream, and has expanded into pig farming and the production of meat products. Although the Quakers continue to be stockholders and active in the factory, it is a throroughly hybrid enterprise now, and the vast majority of producers, workers, and factory administrators are Costa Ricans. In fact, only a handful of Quakers continue to work as dairy farmers.

## Expanding Agriculture and Disappearing Forests

For the first generations of *campesino* settlers and Quakers, the region's montane cloud and rain forests represented a significant resource to be utilized for subsistence living and for their (at least initially) fertile soils after they were cut as an "ecological subsidy" (Griffith et al. ibid.). Until the 1960s and 1970s, there were no roads on which to carry timber to regional markets, so many settlers practiced slash-and-burn agriculture, utilizing wood for construction purposes and often burning what could not be used. As in the rest of the country, burning was most common in lower altitude farms (below nine hundred meters), because at the higher altitudes (above 1,200 meters), this was not always a practical method of clearing lands and raising crops or cattle, due to excessive moisture and cold. It is also for this reason—moisture and cold—that some of the highest altitude lands (above 1,400 meters) have historically remained largely forested. Surplus labor existed in the region practically from the first settlements, and landowners would often rent their lands to laborers, whose task was to clear forest and manage agriculture and livestock. For many settlers, the pressure to make the land *rendir* (surrender, or give) was also motivated to fulfill obligations to private creditors and later state-funded rural credit boards that provided farmers with loans to prepare or expand farms with the expectation that the land's value lay in its "improvements." Until recently, cutting forests has been widely viewed as positive and valuable labor, for its contribution to the creation of human settlements and agricultural production. As one man observed to me, "Forty years ago, the one who cut the most trees was the toughest man, and for a lot of men it was what they did for fun." And "men's work" it generally has been: while girls and adolescent women would work in forests, in foraging, carrying building materials, or even at times assisting in hunting, it has been widely understood that women's ontological closeness to the nonhuman forces of nature made them especially vulnerable in forests, and therefore more suited to agricultural and household tasks.[5]

At the same time, the forests provided a significant resource for subsistence survival, including meat from game, plant foods, construction materials of wood and vines, fuel, and medicinal plants.[6] The forest was also a mysterious and potentially dangerous place, where animals such as jaguars and supernatural forces lurked and terrorized people who would enter it. Even today, residents tell stories of battles that took place between settlers and aggressive jaguars, and draw from a rich body of folklore on forest spirits and creatures to explain unusual happenings.[7] Hunters and *baqueanos* (like José), it is thought, are likely to experience such creatures, since they spend the most time in forests, although younger people increasingly dismiss such stories as superstitions of their grandparents. In addition, it was in the forests where illegal activities such as the production of contraband could be concealed, and they provided a hiding place when authorities came to investigate.

In contrast to ideas of rain forests as unpeopled wildernesses or as the site of an abstract and unpeopled pristine tropical nature, resident Costa Ricans have viewed forests as social spaces marked with human history and agency. Significantly, among *campesinos* of the Monte Verde region, the forests have never been referred to as *selva* (jungle), a term that implies vast expanses of lowland forest wilderness where dangers are heightened and people unwelcome. Rather, they have referred to forests as *montaña,* and more recently *bosque,* words that can be translated as "forest," but also carry the connotations of "woods" in North America and Britain. Like woods, *montaña* and *bosque* represent relatively well-known spaces within which people live, recreate, and travel. So it is that during walks with Manuel in the cloud forest of the Reserva Santa Elena (the story that begins this book), many of his stories were about people who had tried to homestead in the area; where certain incidents took place; and what it was like to travel through the area before it became formally protected.[8] Subsequent to the emergence of the dairy economy in the Monte Verde region throughout the 1960s and 1970s, the improvement of roads and communications between neighbors, and the increased urbanization in Santa Elena and Cerro Plano that has come with tourism, however, few (if any) Monte Verdeans have continued to view the forests as containing the resources for subsistence living.

Ironically, the settlers who relied most directly upon the forests for subsistence living and as supplement to their agricultural pursuits converted the most forests to agricultural land. According to the reports of Costa Rican settlers, by 1950 upwards of 70 percent of forest conversion in the Monte Verde region may have already occurred, especially in lower altitudes of the region (Stuckey 1988: 62), and the last large remaining tracts of continuous forest existed along the Continental Divide at the highest altitudes. This is not to say, however, that no primary forests have existed outside of the area of the Continental Divide, or that *campesinos* have had unchecked ambitions to convert all forest into pastures or agriculture.

For example, agricultural census figures suggest that at least since the early 1970s until the late 1990s, farms in the district of Monte Verde have continued to support important amounts of forest.[9] In the early 1970s, the 128 Monte Verde farms were divided into pastures (51 percent of the area of each farm), cultivars (6 percent), primary forests (36 percent), secondary and regenerating forests (5 percent), and other uses (2 percent, such as buildings) (Ministerio de Economía, Industria y Comercio 1974). By the early 1980s, on the region's 141 farms (average size 28.1 hectares), areas in pasture remained the same as the previous census (51 percent), and cultivars expanded (13.7 percent), presumably at the expense of forested areas (which had decreased to 26 percent primary forest and 3.5 percent secondary growth) (Ministerio de Economía, Industria y Comercio 1987).[10] That is to say, in the early 1970s, the region's farms were on average approximately 40 percent wooded, and in the 1980s they were approximately 30 percent to 35 percent wooded. More recent Ministry of Agriculture figures indicate that the typical Monte Verde farm is 30.5 hectares, in which 46 percent of the area is dedicated to pastures, and 8 percent to cultivars (MAG 1996), leaving up to 46 percent forested (see figure 2.3).[11]

*2.3 Cattle pastures and forest in Santa Elena (photo by author).*

These figures are more meaningful when they are placed in comparative perspective, for it is possible to identify certain patterns, including the fact that landowners in neighboring districts have not integrated forest maintenance into land management strategies to the extent that Monte Verde landowners have. For example, in the early 1970s in the district of La Sierra de Abangares, Guanacaste Province (that borders Santa Elena to the north), farms maintained 8 percent of their area in primary forest and 12 percent in secondary growth, while 66 percent was in pastures (Ministerio de Economía, Industria y Comercio 1974). By 1984, forested areas had been reduced to 3 percent primary forest and 2 percent regenerating growth (Ministerio de Economía, Industria y Comercio 1987). The significant difference in forest maintenance between Monte Verde and this district can be explained partly by the character of land tenure and the economies into which they are integrated. In La Sierra de Abangares, for example, in the early 1980s 60 percent of farms (average size 103 hectares) were owned by *sociedades,* or corporations, including both legal corporations (*sociedades de derecho*) and informal corporations (*sociedades de hecho*). In Monte Verde, by contrast, 78 percent of farms were individually-owned, while 21 percent were owned by sociedades (ibid.).

One reason for differences in rates of forest conversion between these neighboring districts is the distinct uses to which lands have been put. Significantly, La Sierra de Abangares is integrated into the international beef cattle economy, dominated in Costa Rica by *latifundistas* (absentee landowners of large ranches), and supported by Costa Rican subsidies and international development aid funding to produce inexpensive beef for North American consumers (Gudmunson 1983; Edelman 1992). Furthermore, corporate landholders have often been more interested in short-term profits than in preserving forests for their longer-term benefits. In contexts like Monte Verde where small landholdings have predominated and landowners have generally tended to live on their lands or in the communities where they own land, they are often less willing to eliminate forest on the levels of neighboring districts, since forests are important resources for long-term farm management, including watershed and spring protection, wind blockage, construction materials, shade for animals, and sale for timber.

As I will show in the next chapter, Monte Verde environmentalists preferred to purchase land in areas they considered least disturbed by human influence, and therefore many of their efforts were directed to purchasing lands within the Reserva Forestal Arenal. One of the first areas where they focused their land purchase campaigns in the mid 1980s was the Peñas Blancas Valley, which was mainly (92 percent) primary forest. According to one study, "human influence" in the Peñas Blancas Valley was limited to certain identifiable corridors and zones where mainly subsistence homesteads previously existed (Cummings 1989). This study ascertained that out of over eight thousand hectares that were acquired for protection by Monte Verde environmental organizations by 1988, four hundred hectares were "human-created successional habitats." These "disturbed habitats," most of which were created in the 1960s when soils for dairy cattle pasture had reduced fertility in Monte Verde and some dairy and subsistence farmers sought *suelos más frescos* ("fresher soils"), were found mainly along riparian zones and less steeply sloped areas in the Valley. By the late 1980s, these areas were mainly secondary forest and regenerating pastures, and the last cattle were removed in April 1988.

A number of factors ensured that when land purchase campaigns for the Peñas Blancas Valley began in the mid 1980s the land was almost totally closed-canopy forest, the primary forests prized in formal conservation efforts. A major reason for this situation was the inclusion of that area in a national forest reserve related to the Instituto Costarricense de Electricidad's (ICE) Arenal Hydroelectric Dam Project. In the late 1960s, technical studies (i.e., CCT 1968) identified the forested areas south of Arenal, including large areas of the Monte Verde region, as key watersheds for the development of Arenal Lake and the Arenal Hydroelectric Dam, a major development intended to provide electricity for northern Costa Rica, as well as for an irrigation project in the Tempisque lowlands of Guanacaste. When President Daniel Oduber declared the thirty-five thousand-hectare Arenal National Electric Energy Reserve in February 1977 as an absolute protection watershed, landowners within the watershed were angry that they were not consulted, and in response protested legal and processual irregularities related to the law. By April 1977 the size of the reserve was reduced to 18,325 hectares and its name changed to the Reserva Forestal Arenal (ICE 1978). Within the forest reserve, restrictive measures were implemented on the remaining landowners, including the prohibition of bank credits, denial of land titles, prohibition on cutting trees and burning, and abandonment of government support for improving roads and other infrastructure.

The designation of the area as a protected hydrological forest reserve is only partially responsible, however, since people had homesteaded and made claims in the valley since at least the 1940s. As one ex-landowner who owned ten hectares in the valley explained to me, ecological factors, such as the difficulty of access, high precipitation, and insect-borne disease (such as leishmaniasis) had also discouraged large-scale settlement of the valley. Economic factors were also important, including remote access to markets for lumber and dairy products and the increasing intensification of agricultural and dairy production throughout the 1970s in the Monte Verde region. Furthermore, for many of the landowners (including many who lived on the Pacific-slope villages of Monte Verde) their claims in the valley represented second or third landholdings. Some of these were dairy farmers (including Quakers) who planned to use pastures in the perpetually wet valley to feed dairy cattle during the Pacific slope's dry season. Others viewed these lands as a form of insurance if agricultural markets or their own soils worsened, or as a resource to sell at a profit through speculation. In desperate circumstances, they could be worked for family subsistence.

Just as *campesinos* have never been economically homogeneous in Monte Verde, not every landowner or settler has practiced the same styles of preparation and management of farms and forests. For example, until the 1980s, many *campesinos* (especially at lower altitudes in the region) annually prepared pastures by cutting forest edges and burning the non-useful woods and overgrown pastures. Although the Monte Verde region is rich in volcanic soils, variations in soil quality and management (or mismanagement) have affected harvests and pasture quality. Not only did burning in certain areas quickly and efficiently prepare land for agricultural production and pastures, but it also provided an initial pulse of nutrients to the soil that producers would observe in the first agricultural harvests. Another rationale was to clear underbrush and to "heat the earth," since it is believed that agricultural production depends upon "hot" (*caliente*) soils, while forested lands have "cool" (*fresco*) soils. Planting recently cleared or burned land with agriculture has also served (and still serves) as a practice to "soften" hard soils before growing pastures on them.

Not all *campesinos* in the region practiced burning on their land, however. This is particularly true of farmers with lands at the highest altitudes, where ecological conditions meant that farms were generally colder and wetter than those at the lower altitudes, as well as in areas of high rainfall such as San Gerardo and the Peñas Blancas Valley. But some *campesinos* also did not burn out of a recognition of its negative impact upon their soils. For example, one dairy farmer in Santa Elena described his father's land management philosophy and practice:

> Since my dad arrived in the early 1950s, he always encouraged us to save part of our forest and to never burn. He would ask, "Where are you going to get wood for construction and sticks if there's no forest? And what about firewood? What happens to your waters?" He'd read somewhere that burning sterilizes the land. That's why you see that a third of my land is forested today, because we knew what could happen if we were irresponsible. I think that people simply didn't have a good idea of the limitations to the forest that we have today.

Many farmers have cut forests around springs or creeks, which has had the effect of drying them. However, again, not everybody did since some of them understood the significance of water to their pastures, cattle, and agricultural production. "Look," another Santa Elena dairy farmer told me,

> I know the impact that cutting trees has on my waters and pastures, and the impact that it will have on others who live near me. Just look in the lowlands how dry and eroded the soils are because they cut all their trees. I've always thought twice about cutting a tree – it doesn't mean that I won't cut it if it has a use for me, though. It just means I'm careful about where I do it. For example, I will cut a tree so it won't take down other trees, or use one that is fallen already, depending on my needs.

Selecting a tree to cut depends upon its utility for construction or fuel, age, position in relation to other trees and water, and most importantly, species. For example, certain species of *Lauraceae* (laurels), for example, have been in high demand because they are good for house and foundation construction.[12]

Until the 1980s, when state forestry service agents began enforcing laws against cutting trees more rigidly than they ever had, farmers in Monte Verde had a choice: to try to expand into new areas of production at the expense of forests and run the risks of punishment from the state, or to increasingly technify production (Stuckey 1988: 69; Griffith et al. ibid.).[13] In the preparation of their dairy farms, the Quakers discouraged burning, since they recognized that in previously burned areas soil quality was low. They also introduced intensive methods of production for dairy farming, including pasture rotation, artificial insemination, chemical pasture fertilizers, and the use of supplemental feeds (all of which are in common use now). By the 1970s, when agricultural and dairy production decreased as soil fertility declined, Costa Rican farmers also began to adopt semi-intensive and technical methods of production with state support, similar to Quaker methods (Stuckey ibid.: 66). But this was also a period of significant forest conversion outside the immediate area of Santa Elena, because of the expansion of the Monteverde dairy industry. At the same time, instead of automatically cutting more forest, some Costa Rican farmers had sought to rehabilitate the lands that were eroded and unproductive. For example, one Santa Elena woman dairy farmer began reforesting her denuded hillsides with a combination of fruit trees, native species, and exotic species to combat erosion and break powerful winds, long before organized reforestation efforts began.

Strong winds have represented a potent ecological factor affecting production in Monte Verde. One middle-aged Costa Rican dairy farmer observed, "For as long as I can remember in Monte Verde, the winds have been very strong, and that really affects what you do. Of course, there are places where you don't have to worry about it, and some of those are just the place itself, like the orientation of the land, but some are the result of careful planning by the person who left windbreaks." Strong winds from the Atlantic, especially during the dry season, prevent coffee plants from growing well and dairy cattle from high levels of milk production. Some farmers recognized the wind factor in the planning of their farms, leaving forested corridors to act as windbreaks (a theme on which later conservation organization-sponsored reforestation efforts focused). Furthermore, some of these same farmers recognized the importance of not cutting forest on steep lands in order to prevent erosion. Since a highly successful reforestation and windbreak program operated between 1989 and 1994, even more farmers are committed to windbreaks (see chapter 4).

A common assumption among *campesinos* has been that every person has his own level of experience, knowledge, or orientation to farming. One *campesino* from San Luis observed,

> Some agriculturists have a consciousness, a real caring for what they do, as some now say, of loving the earth. They love farming. And their own experience teaches them and gives them wisdom. For example, there are some that would think, "I am going to leave the trees on that knoll, thinking of providing a windbreak or preventing erosion." I think it is natural in them to use the earth in a way that makes it last a long time. They are continuing Creation. Some have done it with fruit trees, and others do it for the waters. They save the springs on their farms, knowing that it is the source of life for their animals and families. This is something you notice at first glance with a lot of farms, that they have protected the forests around their springs. But you should understand that not everyone treats their forests the same way.

There are various ways to interpret the perceived productivity, success, or failure of an individual's methods of farm and forest management, reflecting a diversity of attitudes and practices for the management of land. For example, if a landowner makes a poor decision in choosing what kinds of pastures or crops to plant, experiences declining soil fertility due to burning or overuse of fertilizers, or does not have success in an agricultural endeavor, commentators might point out that the person lacks *conocimiento* (knowledge), which could be the result of a deficit of personal intelligence, experience, formal education, or access to certain technologies. Another especially important factor has been an individual's willingness to work hard to make the land *rendir* (surrender), and poverty or lack of success have often been reckoned in terms of personal unproductivity or laziness (often associated with alcohol abuse) instead of lack of *conocimiento*. Another level of explanation revolves around *la suerte* (luck) and *envidia* (jealousy), in recognition of the supernatural elements that regulate all aspects of an individual's life. Because the land has powers that lie outside of human control, *campesinos* can never necessarily count on getting a return for their hard work, and so agricultural activities have often been spoken of in terms of the "luck," sometimes seen as the dispensation of God, that one has to collect a successful harvest or production levels.[14] *Envidia* can bring *mala suerte* (bad luck), a vindictive form of impact on one's agricultural production or personal success, if a jealous or competitive rival places a supernatural injunction—through witchcraft—upon an individual to decrease his (or his land's) productivity.

The hybridity of *campesino* land management practices is a reflection of the variety of knowledge forms that have shaped *campesino* production and practices of forest conversion. Since at least the 1970s, technology intensive forms of pro-

duction have existed alongside other ways of knowing and making the land productive based on *concimiento popular* ("popular knowledge") (Vargas 1990). For
example, some rural Costa Ricans regulate agricultural activity according to the
phases of the moon, or the *menguante-creciente* complex.[15] This form of knowledge, which regulates planting and harvest times based on cyclical fluctuations in
the moon's gravitational pull, also extends to regulate proper times to cut trees,
depending upon the usage that the person has for that tree. For example, for construction purposes, a tree is more useful (depending upon the species) if it is cut
during *menguante* (waning moon) so that it does not expel all of its internal moisture; if cut during the *creciente* (waxing moon), it expels too much water and can
end up being useless for certain kinds of constructions. If, on the other hand, one
is seeking to collect the sap from a tree (i.e., to produce liquor from the *coyol*
tree), then it is better cut during *creciente*.

But it would be a mistake to totalize this system as the only one that rural
Costa Ricans have used to understand how nature works and agricultural production, since as one middle-aged dairy farmer expressed to me, "Not all
*campesinos* share these ideas. Some yes, they approach things philosophically.
Others no, they simply repeat what their grandparents did, or what someone else
tells them. Some study nature and think about its processes. Others no." To
some extent, this is related to the fact that individuals who own land may engage
in multiple careers during their lives, so that being a farmer is only one productive strategy among others, or a fallback if economic ventures go sour.[16] Moreover, commitments to different knowledge systems are to some extent
differentiated along generational lines, so that many young people in their twen-
ties and thirties claim not to think in terms of popular knowledge and rural traditions, such as the *menguante-creciente* complex. The point is that *campesinos*
have approached agriculture and forest management as a *bricoleurs,* flexibly
adopting and incorporating knowledge from different sources on the basis of
effectiveness, efficiency, and whether or not it makes intellectual sense. As a
result, it has never been possible to say that *campesinos* in Monte Verde represent
a homogeneous social category in terms of the knowledge and practices they
bring to understand and manage forests and agricultural production.

Similarly, there has been no singular Quaker environmental ethic or preservationist attitude toward forests. For example, one of the Quaker settlers explained
that while he decided early on to preserve a large tract of forest on his land, others in the group viewed forests as obstacles to be converted to cattle pastures. He
told me, "We felt very alone in deciding to save our forest. The others believed
that clearing forest was progress." Monteverde residents often joke that one of the
staunchest advocates of forest preservation among the Quakers was during the
early years virtually "a chain saw company representative," due to his enthusiasm
for cutting trees and teaching *campesinos* how to use chain saws. Furthermore,

throughout the 1950s and 1960s, there was no singular Quaker interest in protecting the watershed from the encroachments of *campesinos* who intended to establish homesteads or gardens in the watershed. As one long-time resident of Monteverde observed, "When it came to getting those squatters out and protecting the watershed, it really depended on the family that owned the land. The rest of the people were busy doing other things, like dairy farming, or simply not interested in constantly going in to roust squatters." Like Costa Rican attitudes toward forests, Quaker attitudes have not been static, and a number of Quakers who initially saw deforestation as progress later became central actors in promoting the formal protection of the watershed and the establishment of a forest preserve on their lands. But this image—of Quakers as stauch environmental activists—is the most persistent image of them. The fact that preservationist convictions evolved over time, or that even today not all Quakers share equally in their commitments to conservation, is rarely mentioned.

The important issue here is that forest conversion and agricultural expansion in the Monte Verde region has been related to a number of factors, including national policies that have recognized land ownership on the basis of clearing forest, the creation of economic opportunities with access to new markets (especially the dairy industry), credits and loans that have been extended to *campesinos* with the expectation of "improvements" to the land, technical assistance from the government and the dairy cooperative, ecological and climatological conditions, attitudes toward forests and production, and the fact that people have depended upon agricultural lands to survive. Forests have not been converted necessarily out of an ignorance of the consequences. On the contrary, arguments that *campesinos* have been inherently destructive toward forests overlook the specific conditions under which these people have both converted forest, and *not* converted forest, to agricultural lands.

## The Neoliberal Crunch

When environmental groups began purchasing lands for formal protection in earnest during the mid 1980s, landowners also had to be willing to sell their lands, because the declaration of those lands as "frozen" for development alone did not expropriate them. For José, whose story opened this section of the book, the reason for selling his land was his brother's indebtedness, which of course raises important questions about the diminishing returns of small-scale farming in the Monte Verde region for certain landowners during the late 1980s and early 1990s. The political and economic vulnerability of small-scale farmers was related to broad structural changes in the Costa Rican agrarian situation since the 1970s, including the transition from intensive state involvement in the countryside to neoliberal conditions of reduced state mediation of the agricultural economy.

Throughout much of the latter half of the twentieth century, the Costa Rican state pursued welfare and reformist policies in social and economic arenas, to the extent that state agencies and regulations eventually permeated virtually every aspect of economic production and development (Rovira Mas 1989; C. Rodriguez 1993; Edelman 1999). Particularly after the 1948 civil war, state interventionism satisfied several demands, including social stabilization and the legitimation of the new post–civil war political order (Rojas Bolaños 1990). In agrarian sectors, these included price supports and production subsidies, funds for rural credit boards, the creation of infrastructure, and technical assistance. But world coffee prices plummeted in 1978, and in 1979 oil prices rose abruptly and the Nicaraguan Revolution interrupted overland traffic and commerce. These factors, as well as recession in other countries, negatively affected foreign demand for Costa Rican exports, all of which contributed to rising interest rates and an overvalued currency. The *colón*, which had been fixed to the dollar, was allowed to float in 1980, and by the end of the year, its value had nearly halved (Edelman and Kenen 1989: 188). During the "debt crisis" that ensued, inflation and underemployment rates soared, and the decline of real wages became the reality for the majority of Costa Ricans. Large public sector deficits forced the acceptance of three rounds of IMF structural adjustment measures and the aggressive pursuit of free-market economic policies throughout the 1980s and 1990s. Structural adjustment marked the rupture of previous forms of social consensus achieved and maintained through mediating state institutions, and the introduction of new forms of economic development and insertion into international markets (Trejos 1990; Korten 1997). This period also saw the expansion of the nontraditional agroexport sector of ornamental plants and tropical fruits, as well as the increasing politicization and organizing efforts of *campesinos* among the public sectors rejecting these fundamental changes in state policy (Edelman 1999).

In Monte Verde, where the state's direct presence has historically been relatively minor compared with other parts of the country, medium and small-scale farmers have nonetheless been profoundly impacted by the retraction of state mediation in the public agricultural sector. In response to the debt crisis during the 1980s, extensionists from the Ministry of Agriculture and Livestock (MAG) and the Monteverde dairy cooperative encouraged farmers to "de-technify" their production processes. That is, they were counseled to not fertilize their pastures and feed cattle only grass in hopes that production costs would fall more than their income (Stuckey ibid.: 66). As a result, milk production decreased, and some farmers faced with trying to make ends meet began to look at alternatives beyond dairy farming, including selling their lands. Further, the opening up of national markets to international competition for staples like corn, beans, and rice that *campesinos* had traditionally produced meant that small-scale producers could no longer necessarily find markets for their production.

The height of the neoliberal transition and the inability of many landowners to make a living on their lands through traditional agriculture also coincided with the rise of Costa Rica as a safe and peaceful tourist destination, and environmentalist land purchase campaigns based on international donations during the 1980s and 1990s. The limits the Costa Rican state had placed on supporting and expanding traditional agricultural sectors during this period were less the product of a unified ideological commitment to conservation than they were constraints placed on government action and investment by structural adjustment programs. Even during periods of structural adjustment, the Costa Rican state has maintained a contradictory position with respect to the environment, favoring the ecologically destructive nontraditional agroexports to earn foreign exchange, while at the same time supporting the expansion of conservation areas and policies (Rodriguez 1994). The combination of declining state support for expanding traditional agricultural production meant that many *campesinos* and landowners found themselves stuck in a crunch. They could not work their lands in undeveloped areas because, as a result of its own fiscal crises, the government would not fund the expansion of agriculture as it had for decades by providing low-interest loans and credits, road building, and the construction of other infrastructure. Moreover, it would not expropriate most of the landowners whose lands fell within the hydrological reserve. Monte Verde *campesinos* did not take these developments passively, and joined together with national organizations such as UPANacional (National Union of Small and Medium Agricultural Producers) to protest the government's new policies and pressure it to find new forms of public support for rural communities and and small-scale market agriculture (see chapter 5; Edelman 1999). But the erosion of the state's mediating and dominating role in rural communities – even if Monte Verde had never been a region of intensive state involvement – meant that environmental activists found that in many cases, there were willing sellers of land in the areas they most desired, that is closed-canopy wilderness.

# Notes

1. Before environmentalist land purchase campaigns, there were two efforts to systematize land ownership. The first took place in the 1950s, with results widely considered disastrous (people claim the land agent took bribes to make claims larger), and the second took place in the 1960s through the national Instituto de Tierras y Colonización (ITCO), or Institute of Lands and Colonization.

2. For example, by the 1920s, the village of Turín, 12 kilometers from Santa Elena, had a large number of coffee farms and a pool of migrant labor from the Meseta Central. In the 1920s, the church was able to procure an expensive set of bells from Italy, a reflection of its relative prosperity.

3. As of the mid 1990s, one man in Los Tornos lived from gold mining on his land.

4. Costa Ricans did not actually become shareholders until 1976. After an internal debate and consensus-building, the Cheese Plant offered milk producers reduced-rate shares (Howard ibid.).

5. Nygren (1993) finds a similar set of beliefs near Turrialba. Although fewer families live in the midst of forest, these ideas persist in the belief that large cats such as pumas are especially attracted to women.

6. Common sources of meat included *danta* (tapir), *saíno* (peccary), *venado* (deer), *tepezcuintle* (agouti), *cosuco* (armadillo), *pavón* and *pava negra* (black guan), and to a lesser extent opossum, *mapache* (racoon), *pizote* (coatimundi), *mono* (different species of monkey), and large cats (puma, jaguars, ocelots, and so on). Common forest foods included *aguacate del monte* (mountain avocado), *palmito* (hearts of palm), *zapote,* and *papamiel.* Common sources of wood for construction have included *cedro* (cedar), *burío,* species from the lauraceae family, *tempisque, danto, bejuco del hombre,* and *frijolillo* (vines). Common fuel woods included *níspero lechoso, las murtas, chancho blanco, lagarto amarillo, ratoncillo,* and *fosforillo.* Guarumo (*Cecropia*), whose leaves had been used in a tea for weight loss, was commonly used to create canals for the production of contraband.

    Monte Verdeans never developed self-sufficiency in treating diseases and accidents, but would take seriously sick people to medical doctors in Las Juntas, Puntarenas and to the Monteverde Quaker "doctor" who administered medicines after the early 1950s. Nevertheless, even with the establishment of the state health clinic in Santa Elena, many (especially) older people have relied on commonly available medicinal plants. In most cases, these plants are maintained in house gardens or are available in pastures and *charral.* This information is based on interviews and is complemented by Vargas (1990) and Gutierrez (1983).

7. This folklore includes stories of el Dueño del Monte (Owner of the Forest), *brujos* (witches), La Llorona (the Weeping Woman), and *duendes* (trolls), among others.

8. Other categories of forest and vegetational cover emphasize the landscape's regenerative power: *tacotal* (more than five years of growth, often called "secondary forest"), *charral* (one to two years of growth after a forest is cut), and *monte* (weeds).

9. I present the following figures only to give a *very general* indication of the relationship between farmers and forest management in the area, since agricultural census figures can be notoriously inaccurate. For example, farmers have been reluctant to participate fully and honestly in censuses, believing that they represent an opportunity for the government to calculate taxes. Furthermore, the categories utilized in censuses have tended to obscure precise differentiations between farms and the practices of their managers. Censuses also do not necessarily explain changes in measurement criteria between censuses or changes in administrative partitioning (for example, between the 1963 census—Ministerio de Economía, Industria y Comercio 1965— and the 1973 census, the district of Monte Verde was created. As a result, the 1963 census figures do not show the specificity that later censuses do; therefore I have not included these figures

in this discussion). Another possible source for these indications would be aerial photographs, although they may not show different types of forest as clearly.

10. The difference in cultivar figures is probably related to the Costa Rican debt crisis of the early 1980s. During this period, dairy prices decreased severely and some dairy farmers sought alternatives to dairy production, including cultivar production. One study conducted in the early 1980s that examines seven dairy farms in the Monte Verde region offers a different calculation of land use than the 1984 census (Gutierrez 1983). In this study, 44 percent of the area of each farm was dedicated to pastures, and 35.7 percent dedicated to primary forest (ibid.: 35). Gutierrez suggests that the relatively high percentage of forests were in windy parts of the farms and the result of farmer awareness generated by the presence of a forest preserve (Monteverde Cloud Forest Preserve).

11. This 8 percent also includes coffee production, which in Monte Verde has traditionally been small scale and complementary to dairy production. In the last several years there has been an increase in the number of farms (currently over eighty) and amount of farm area dedicated to coffee production (on average one hectare), related to the expansion of the Santa Elena Agricultural Cooperative's efforts at processing and placing Monte Verde coffee in national and international markets.

12. Biologists point out that the older the tree, the more important it is for seed dispersal, and therefore more important to keep standing. Until recently, when biologists began pointing these facts out to them, this has not been of great concern for *campesinos,* who are usually more interested in the tree's utility for themselves than for birds or for reproduction of the tree.

13. In 1969, the Costa Rican government passed the first major Ley Forestal, intended to regulate the cutting of forest in *tierras baldías.* Enforcement of this law throughout the 1970s and 1980s was patchy, at best, even though a forest ranger was posted in Santa Elena.

14. Gudeman and Rivera (1990: 26) suggest that this is a common factor across Latin America.

15. The period of *creciente* (waxing moon) corresponds to a time to plant root crops (such as potatoes), while *menguante* (waning moon) corresponds to a time to grow tall plants (such as corn). During the *creciente,* the moon's pull is strong, and a tall growing plant will grow too tall if planted during this period and not produce sufficient harvest and is susceptible to strong winds; if planted during a *menguante* period, a tall growing plant will not grow tall, but will grow short and strong to withstand strong winds. This system extends as well to human and animal bodies, so that blood is understood to flow stronger in *creciente* than *menguante* (and therefore has application to castration of animals, cutting hair, etc.). Nygren (1993) observes similar forms of knowledge in Monte.

16. One of my next-door neighbors from Santa Elena is a good example of this multi-career individual. He has owned land in the Peñas Blancas Valley, Santa Elena, San Gerardo, the Montes de los Olivos region, and San Carlos. Now in his sixties, he has worked in many jobs, including dairy farming, renting his lands, cultivating vegetables for markets, selling clothing, owning a sawmill and lumber business, and lately working as a restauranteur, among many other jobs.

# 3 Uneven Terrain: The Practice and Politics of "Saving" Monte Verde

"So where is this famous tropical deforestation that we hear so much about in the U.S.?"

"Everywhere. This was once a continuous rain forest. The Costa Ricans cut it down to create farmland. That's why they saved the forests in Monte Verde."

Exchange between a North American tourist and a North American biology student, on the bus to Monte Verde, mid 1990s

Brief as it is, this exchange I overheard on the public bus to Monte Verde illustrates one of the central reasons environmental activists have given for their efforts to formally protect Monte Verde cloud and rain forests. That is, the threat to endangered habitats and species posed by destructive Costa Rican land management practices. Motivating the question is the premise that deforestation has to be happening in order to be fully appreciated, since the tourist would not have asked if he saw it in front of him. It is an attitude at least partially conditioned by North American and European media images in which tropical forests are commonly represented as being cut and burned. This idea is confirmed by what the tourist told me: such media images have always depressed him, so he came to Costa Rica to see what tropical forest loss is like first hand.

The fact that the conversion of forests to farmland along this road happened many decades before our bus ride did not seem to disappoint him. But my interactions with him suggest that there is extraordinary power in these images of destruction and the assumptions they make about global processes, motivating people to travel to remote places to experience tropical nature before its presumed disappearance. Significantly, these images are often vague about their geographical and historical reference, and so specific landscapes become stand-ins for a broader category, as in Monte Verde standing in for tropical forests elsewhere. They apparently were for this tourist, who came to this part of Costa Rica with no sense of its particular landscape history. In fact, as this chapter will discuss, such conflations of universal and specific have served as intellectual justification in efforts to formally protect Monte Verde landscapes.

That the exchange took place on the famously unpaved road to Monte Verde is interesting, because the landscape along the road has triggered actions to formally protect cloud and rain forests in the highlands, based on the premise that Costa Ricans have not properly managed the landscape. For some travelers, the adventuresome experience of the steep climb from the Pacific coastal lowlands to Monte Verde is breathtaking, indeed, one of the key attractions of going there (even while locals along the road clamor for it to be paved). Travelers are rewarded with spectacular views of rolling hills and the Pacific lowlands, as well as picturesque rural landscapes of cattle ranches, homesteads surrounded by fruit trees, and sturdy men with machetes on horseback. But for other visitors and Monte Verde residents, the destruction of tropical nature confronts them every time they drive on this road, for what is pleasant and photogenic rural countryside to one person represents biologically sterile, ecologically degraded, and aesthetically unpleasing landscape to another. Especially in comparison to the exceptional ecological complexities of the cloud and rain forests that many people come to Monte Verde to interact with, such countryside may be not particularly interesting, and even deeply troubling (see figure 3.1).

*3.1. On the road to Monte Verde (photo by author).*

One North American biologist, with long-term research and fundraising connections to Monte Verde, once told me about the profound disquiet she has felt traveling along that road:

> I ride the Monte Verde bus a lot, and I've worked in the Monteverde forest, the Cloud Forest Preserve, so many hundreds of days that it's like a part of my body.... I had begun teaching and I found myself developing from my general ecology course a lecture on human population growth, and I would give a sort of Peter Raven-type lecture about deforestation and I would write and give these good lectures about deforestation.[1] But at the same time I was in denial ... and I would come home and feel totally out of whack with myself, until I finally realized these were pretty awful facts that I was telling people. ... One day I was riding the bus down from Monteverde. I think I was on my way home and I looked at that deforested landscape and it's just still so emotional for me. I look at that and I thought, this is happening to the whole globe! It's so overwhelming. How can anybody do anything? It's just so overwhelming. I, my students are going to inherit a world worse than the one we're in. Everything's so big, complex, so connected, so global. I can't do anything. And then suddenly something in my brain or heart or soul, I don't know which, said "Shut up and get to work!"

What followed was this woman's intense commitment to educate people in North America and Europe on the wonders and disappearance of the world's tropical rain forests, and the creation of an organization dedicated to raising funds to sponsor land purchases and other conservation projects. By the late 1990s, her organization channeled as much as $500,000 for a Monte Verde environmental group, as well as several hundred thousand dollars for nongovernmental organizations in other parts of Latin America and the northeastern U.S. Like other activists who raised money to buy Monte Verde area forests, she both inspired and rode a wave of interest in North America, Europe, and Japan, especially among children, of "saving the rain forest." The majority of that money was applied to the purchase of lands for growing forest preserves, in adopt-an-acre programs in which donors received a certificate of gratitude noting how many acres their donation helped to purchase.

Although they more or less ended by the early 1990s, before I began long-term research there, internationally funded land purchase campaigns have been Monte Verde's most important and transformative encounter with environmentalism (see figure 3.2). This chapter is about these efforts to formally protect Monte Verde's cloud and rain forests, focusing on how scientists, environmental activists, nongovernmental organizations, and international donors defined what areas are relevant for formal protection, and why the establishment of forest preserves was promoted as a central solution to regional problems of environmental degradation. The organization and support for these efforts is transnational and transcultural, drawing from governmental, scientific, private nongovernmental, and grassroots sources in over forty countries. The vast majority of the literally hundreds of thousands of people who helped "save" Monte Verde forests have never been there, seen the forests (except maybe in a television documentary, or a few photographs in a magazine or slide show), or met any of the people who live there. This chapter asks, how did environmental activists working in Monte Verde channel the vast financial resources and moral authority of culturally and politically diverse people in their efforts to implement solutions to problems of environmental degradation? What impacts did this have on how residents perceive and experience the relationship between themselves and the forests near and in which they live?

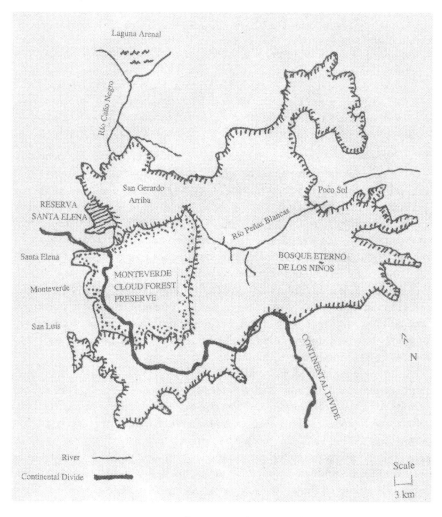

*3.2. Map of Monte Verde Protected Areas.*

Establishing and protecting forest preserves is only one strategy among several that Monte Verde activists and organizations have employed in their efforts to conserve tropical nature. But the creation of forest preserves reflects specific acts of inclusion, negation, and denial, and crystallizes in space certain ideologies of human-nature separation by creating new legal and conceptual designations (from restricted-use areas to buffer zones) that imply certain kinds of interventions and controls placed on peoples' activities. Moreover, land purchases have created the conditions for the expansion of the tourism industry in Monte Verde; clarified, and in some cases even established for the first time, records of land ownership; and provoked profound conflict because of the controversial assumptions they make about rural Costa Rican relations with the landscape. The purchase of lands enabled the formation and helped define the legitimacy (or illegitimacy) of environmental organizations, helping situate these groups as new public-sector authorities with which more traditional public authorities had to coordinate on a number of key issues of regional concern, such as road maintenance, law enforcement, and visions for future economic development. As a technical, but also profoundly political, solution to problems of environmental degradation, land purchase campaigns played a key role in the effective centralization of Monte Verde conservation efforts within the organizations that own or buy forested land, and have had important consequences for subsequent environmentalist action. Therefore, it is relevant to ask how certain areas became targets of environmentalist interventions, for whose benefit, and upon what symbolic, material, and conceptual foundations such practices have been based.

Furthermore, it is necessary to assess the multiple and shifting intellectual arguments environmentalists have used in their justifications to "save" Monte Verde's cloud and rain forests, and more importantly, to explore the profoundly uneven relationship between these abstract concerns and the complex realities of land purchase campaigns. Under this kind of scrutiny, it is difficult to specify any single or privileged logic underlying preserve design. Indeed, because it connotes rational agency, "design" may be an inadequate term to make sense of these efforts, drawing as they did on the inconsistent and shifting status of scientific knowledge and expertise in conservation; ideological skepticism over state and public forms of land tenure and management; real estate market opportunism; a fragmented and competitive field of non-governmental organizations; the influential concerns of international donors; stereotypes about rural Costa Ricans; powerful media representations of tropical forests circulating in Northern countries; ongoing political-economic changes in the countryside that in certain circumstances favored the interests of environmentalists over farmers; and the vigorous critical reactions of some Monte Verdeans to the land purchase campaigns themselves. In order to draw out the subtleties raised by these issues, this chapter is divided into three sections: Part One considers the dynamic and multiple knowledges and arguments used to justify the formal protection of tropical nature in Monte Verde; part two examines the actual land purchase campaigns; and part three examines the consequences of the land purchase campaigns, for residents of the Monte Verde region and for the institutions that purchased land.

## Part One: Practicing Science and Protecting Nature

In his book on the history of nature conservation in Costa Rica, the biologist Luis Fournier observes that "[An] aspect that has been of great importance for the strength of the Costa Rican conservation movement is that of basic research in the natural and geological sciences, a process in which ... many Costa Rican and international scientists and naturalists have participated" (1991: 100, my translation; cf. Gómez and Savage 1983). Since the beginnings of the Monte Verde environmental movement, scientists have also "stood in for nature" as its advocate (Yearly 1993). Indeed, Monte Verde's first protected forest was the direct result of one scientist's commitment to conservation, and later conservation efforts are saturated with scientific influences (see Part Two, below). Because of Monte Verde's high profile in international environmentalist circles, scientific research there can likewise gain a high international profile if it has conservation implications, as a recent article in *Nature* on the links between Monte Verde amphibian disappearance and climate change demonstrates (Pounds et al. 1999).

The cloud forests around Monte Verde are among the most closely studied in the ecological literature, a direct result of the relatively high concentration and prestige of researchers who work there. In the thirty-year period between 1966 and 1996, 253 scientific articles were published on Monte Verde (Nadkarni 2000: 12). Several dozen scientists are in residence or visiting at any one point. The majority are North American ecologists, botanists, herpetologists and ornithologists, who have typically worked in both the core protected areas and private lands in buffer zones, and lived in or near the mainly North American village of Monteverde. The Quaker settlers invited the first North American scientists in the early and mid 1960s, who were impressed by the unique cloud forests and endemic species like the golden toad, which were largely unknown to science at the time. After the first steps toward formal protection were made in the early 1970s, the Organization for Tropical Studies brought graduate ecology courses, and many students dedicated their careers to cloud forests based on this experience.[2] Several U.S.-funded projects in the 1980s brought scientists for pollination and phenological studies, adding to Monteverde's profile in the discipline. The "biological grapevine" (disciplinary gossip) has emphasized the accommodating Quaker community and common English language, helping establish Monteverde as a friendly place to do research. As a North American ecologist who holds an academic appointment in the U.S. and owns a house in Monteverde village explained: "One reason we liked it is that there would be life outside of research. It's a rural community. There are square dances, our kids could grow up here during our summer breaks."[3]

Despite the concentration of scientists, their direct involvement in Monte Verde nature protection efforts has been uneven. For some researchers, there is little (if any) direct connection between their specific research and the concerns of environmental activism or of "saving" Monte Verde's forests and biodiversity. For others, however, there is a persistent struggle to ensure the relevance of their research to environmental practice and policymaking. Researchers, including those who consider themselves doing pure research, often express frustration that the information they produce about ecosystem dynamics, flora, and fauna does not seem to be taken into account in the planning and elaboration of environmental organization agendas. To some extent this is because there is no singular scientific community or authority that exerts a consistent collective or institutional influence (even though the Tropical Science Center is a key conservation institution; see Part Two), and because researchers often move into the area for temporary periods. Furthermore, while they may share their findings with each other on an informal basis or with ecology student groups, there is no central research facility or repository for information.[4] This fact recently encouraged two ecologists with long-standing research ties to the region to publish a massive tome—*Monteverde: Ecology and Conservation of a Tropical Cloud Forest* (Nadkarni and Wheelwright 2000)—that collects a wide variety and large number of scientific articles.[5]

Certain general concepts have provided a cachet of scientific validity in nature protection efforts, including the concept of ecosystem, and more recently, reflecting changes in ecology itself, biodiversity. Introduced in the 1930s, the ecosystem concept serves as an analogy for the study of the interconnections and relationships between organisms within a "natural community" without reducing the community to a single species or organism (Hagen 1992; Pickett et al. 1992). As a discipline, ecosystem ecology has historically focused on describing the structure and movement of energy (biomass, nutrients, and so on) through a microcosmic organic field. This perspective, which constitutes nature as internally systemic (and in various formations as balanced and equilibrant), has been highly influential in the historical and contemporary construction of environmentalist policies, providing an apparently rational basis for the establishment and management of protected nature preserves (McIntosh 1985; Chase 1987; Reid and Miller 1989; Shrader-Frechette and McCoy 1993; Budiansky 1995). When nature is viewed as a place where stability and holism are salient (or, in the language of equilibrium, in a state of climax), the purpose of nature protection and management is to preserve the systemic interactions that already exist, which has tended to place special value on closed-canopy forests as indications of forest ecosystems in equilibrium and climax (Leach and Mearns 1996a: 10). Although ecology as a discipline claims to cover all living organisms (including humans) (Owen 1974), ecosystemic approaches have tended to exclude people and their activities from the world of nature, generally viewing nature as a habitat or wilderness, and human involvement in it as anthropogenic intervention or "disturbance" (Adams and McShane 1992; Gómez-Pompa and Kaus 1992; Bonner 1993; Shrader-Frechette and McCoy 1995; Cronon ibid.; Guha 1997a). Indeed, this ecosystem model is a constitutive orientation in Monte Verde's land purchase campaigns.

However, as methodological and theoretical changes have occurred within ecology in recent decades, growing numbers of ecologists have rejected the ecosystem concept's holism and equilibrium-bias in favor of more dynamic and complex models, and this is true in Monte Verde. Population (also called evolutionary) ecology has offered an alternative to the atheoretical and overly functionalistic tendencies of ecosystem ecology. In population-evolutionary approaches, the focus is upon the patterns and causes of change in the distribution and abundance of organisms in space and time, and based on the generation of quantitative and predictive theories, as opposed to the originally descriptive concerns of ecosystem ecologists (Peters 1991; Pickett et al. 1994). At the center of this approach are the genetical dynamics of species populations and their evolution in relationship with certain biotic and abiotic factors. Population ecologists have argued that the chances of extinction increase as species population sizes decline and as populations become more isolated (Pimm 1991), creating a dilemma for environmental policymaking based on ecosystems philosophy: establishing bounded nature preserves can jeopardize the genetic fitness of certain populations of flora and fauna, particularly migratory species or megafauna that require large spaces in which to move.[6] Reconceived from this perspective, formal nature protection emphasizes the prevention of species extinction and the maintenance of the evolutionary fitness (avoidance of debilitating effects of inbreeding) of specific populations and the ecological integrity of natural communities ("stability" through species diversity). Because natural systems are open and lack a stable equilibrium (and are often fragmented by both natural and human actions), it is impossible to establish natural areas that are sufficiently large to be self-contained because species will move in and out of surrounding human communities (McNaughton 1989).

Reflecting the new complexities this raises for nature protection, one Monte Verde scientist once told me, "Absolute protection is more a philosophical than a biological proposition." Focusing on genetic interactions complicates any notion of protecting nature by bounding it, forcing a rethinking of preserve design to provide the most certain form of protection for the greatest number and quality of species and interactions. It requires seeing nature not so much as a limited spatial quantity or boundable place, but as any number of interactive processes that do not respect the arbitrary boundaries of forest preserves.

As a result, one ecologist explained to me that nowadays it is more appropriate to think of Monte Verde conservationists as being "in the business of biodiversity." As a metaphor for the diversity of (human and nonhuman) species and the ecological interactions that sustain them, "biodiversity" is a deliberately value-laden concept that seeks both to explain nature and to change human relationships to it by recognizing their mutual articulation (Wilson 1988). According to Takacs, the rise of the term biodiversity since the mid 1980s signals the fact that "Biologists hope to have a say in forging a new ethics, new moral codes, even new faiths. By staking out new sources of power for themselves, they ultimately hope to gain control over nature, specifically over how and where and even why wild organisms and natural processes are allowed to endure. By altering our mental configurations of nature, biologists seek to alter the geographical configurations of nature" (1996: 2). It is worth noting that biodiversity offers a physical and epistemological reorganization of space, bringing new social populations into visibility and managerial focus. Because biodiversity exists beyond formally protected preserves, it blurs the boundaries between human and biotic communities by spatially expanding what counts as natural, justifying interventions in rural economies and communities to protect ecological interactions.

Some of these concerns have been reflected in conservation biology, a "crisis discipline" with several practitioners in Monte Verde (Wheelwright 2000), which seeks to apply rational principles to preserving and managing nature threatened by human activities (Soulé 1985; Western and Pearl 1989; C. Carroll 1992: 348). Drawing heavily on population dynamic theories, its practitioners simultaneously emphasize the unboundedness and instability of biotic communities and ecosystems but organize their interventions around deliberately inducing a change in human relationships to those systems. Since the late 1980s, conservation biologists in Monte Verde have sought to shift attention from expanding core (bounded and patrolled) preserve areas to the protection and regeneration of natural processes outside of core forest preserves (Wheelwright ibid.). Forest patches left in working landscapes may maintain species and interactions, and because changes in moisture and temperature along an altitudinal gradient result in high species diversity, many species are left out of large protected areas (Musinsky 1991; Powell and Bjork 1993; Guindon 1996: 168–9). For the most part, these efforts to preserve small forest patches and create forest corridors in Monte Verde have been organized around the charismatic image of the resplendent quetzal, and to a lesser extent, other frugivorous birds like toucans and bellbirds (Guindon 2000).

But there is also epistemological dynamism at the heart of scientific concepts when they are applied to the practice of formal landscape and species protection and management. This dynamism is partially related to the fact that coming from disciplines with distinctive and shifting research foci and methodologies, ecologists (across subdisciplines such as physiological, population, community, ecosystem, behavioral, and landscape ecologies), botanists, herpetologists, ornithologists, and so on are not necessarily in agreement about how tropical nature works and on what level or scale to orient protection efforts and management techniques. Because concepts like natural community, biological corridor, ecosystem, and so on, have differential and debatable meanings across and within these subfields, there is no single criterion to decide at what level or scale to devote protection and management (Perlman and Adelson 1997). Moreover, technically motivated questions such as how big a space must be to ensure protection inevitably lead to questions about what species or species interactions to focus on, what is acceptable to exclude from protection, and how to define the system's "stability" (of which there are many definitions; see McCoy and Shrader-Frechette 1992). These may offer provocative discussions that incorporate the latest theoretical developments but the potential for technical debate and the general urgency of finding solutions has often undermined their mobilization, especially on a scale as large as has occurred in Monte Verde protection efforts (Pimm 1991; Shrader-Frechette and McCoy ibid.; Harper 1992; Mann and Plummer 1995; Perlman and Adelson ibid.).

So although they may as individuals have sophisticated technical perspectives on science as it applies to conservation, Monte Verde scientists often rely less upon technical concepts, language, or justifications for nature protection strategies than on utilitarian and ethical explanations. For example, one biologist who has done long-term research there emphasizes that biologists generally have profound emotional connections to their study sites, a condition that predisposes them to become active in formal protection efforts: "Because biologists are nature lovers, we're doing something we really enjoy. We work in beautiful places and we love nature, that's why we're doing it, and we don't want to see it disappear." In addition, there is a practical concern that for many the forests represent the primary and necessary grounds upon which field research is conducted. The same individual added that this is especially important if a scientist has aspirations of long-term study in one particular place: "If you're doing field biology, especially if it's got some of the long term aspects to it, man, not having your study site turned into a cornfield or a cacao plantation really counts. So having preserved land, and having it preserved in perpetuity, can mean a lot in long term studies. So a lot of biologists involved in conservation feel very practical, personal considerations regarding forests and the relationship to their research." A unifying factor among biologists has therefore been the desire to neutralize threats to the forest first, mainly by promoting the establishment of forest preserves and biological corridors.

Scientists openly recognize that they know little about the forests and species environmental activism is trying to protect and manage. In fact, it is this very lack of knowledge that justifies the formal protection of tropical forests in the first place. According to one resident ecologist, this is partly related to the belief that people need information on what is in the forests before they can decide what to do with them: "Maybe I'm biased [laughs], but biology gives you a certain perspective. It makes you think about and try to know what it is you are going to protect.... If people have the information, if they know what is there, they can decide to clear cut it all, but at least they know what they are doing and what they'll lose." Furthermore, because of the widespread loss of habitat, problems of deforestation require urgent actions. The same ecologist continued that the currency of scientific knowledge is irrelevant because:

> We cannot simply say that we will wait until there is enough evidence to prove to everyone else that rain forest species cure cancer or that these forests serve as global carbon sinks or how these forests specifically serve as watersheds for the lowlands. All these reasons are important, but we have to take action to save these forests before they get destroyed, because of the incredible potential of these forests to serve all of humanity. Sure, we do not know everything that is in those forests when we buy them, but we would be fiddling while Rome burns if we did not make an effort to save them now.

As a result, some have argued that "academic" discussions over conservation obscure the necessity to act to formally protect nature in the first place, or as one prominent North American biologist who works in northwestern Costa Rica has famously said, "The challenge now is to get the academic community to stop intellectualizing conservation to death and get out there and actually do something" (Austin 1990). Armed with such arguments—that forest species can potentially hold the cures to diseases like cancer or solutions to problems of global warming—activists and scientists can assert that the continued conversion of tropical forests or the criticism of actions to prevent it become immoral acts (Janzen 1986; Allen 1988).

As the next sections will show, over time scientific justifications and approaches to nature protection have changed in Monte Verde. Early efforts (1970s and 1980s) were largely dedicated to establishing and expanding preserves, while later efforts (1990s and 2000s) have been more oriented toward what is happening outside the core protected areas. But in important respects the two trends overlap and interpenetrate (even today), and this is precisely because more than scientific concerns have been driving the establishment of protected areas and land purchase campaigns. In the next section, I turn to an examination of how a number of processes and discourses converged over the years—international funding, structural changes in rural economies, representations of rural Costa Ricans in widely circulating discourses, and so on—that undermine any clear sense of rational conservation design and management.

## Part Two: Establishing Monte Verde's Formally Protected Areas

When it was founded in 1972, the Monteverde Cloud Forest Preserve (MCFP) became Costa Rica's first formally protected cloud forest. The story of its founding is now well-known in environmentalist and ecotourist literature, which usually emphasizes that the Quakers took the first, albeit informal, steps toward forest protection by formalizing their 554-hectare watershed as a preserve. It was actually a North American ecology graduate student studying mixed-feeding bird flocks who motivated and arranged the creation of the preserve. The graduate student became concerned over the threat of encroaching colonization in and near the Quaker watershed, and purchased several adjacent homesteads and land claims with the intention of preserving the ecosystems and their unique species in perpetuity. He was also able to secure a promise from the Guacimal Land Company (the company from which the Quakers bought their original land claims in 1951) to donate 328 of its remaining hectares in the area, on the condition that he would find or establish a legal organization to take over the land's administration (Tosi 1992: 2). In 1973 when he did not secure the interest or participation of the Quakers or the Organization for Tropical Studies, he approached the forestry consulting organization Tropical Science Center (TSC; also known by its Spanish name, Centro Científico Tropical, or CCT) based in San José.[7] A TSC founder explains why his institution was interested:

> At that time there were almost no national parks and the TSC had a program for the creation of private reserves for biological research and teaching, each representative of the different ecological zones of the country. It interested us immediately. Then, he [the graduate student] brought us to make an inspection of the lands he had been offered, which are found in the area of dwarf forest known today as "Brillantes." As was expected, we agreed that the area merited protection, but we explained to [the graduate student] that to establish a Costa Rican legal organization required various founding members, sufficient money, and months or possibly years of legal transactions. Since the offer was conditional upon an almost immediate acceptance, we mentioned that our organization fulfilled all the requirements to receive the donation. That was then how we accepted [the graduate student] as a member and the TSC initiated the transactions, that finished in April 1973 with the acquisition of the 328 hectares in Brillantes. It cost us one *colón* (Tosi 1992: 2; my translation).

During the late 1960s, the TSC had undertaken a study of land use capacity of the Peñas Blancas watershed (on the Atlantic side of the continental divide) on behalf of the government's National Planification Office (OFIPLAN). TSC identified the area, including the whole Cordillera de Tilarán, as a crucial watershed for the rivers that fed into the proposed Arenal hydroelectric dam project of the Instituto Costarricense de Electricidad (CCT 1968).[8] In their report, TSC classified the lands as "without economic use" and recommended that they be protected as forest preserves.[9]

By supporting the Monteverde Cloud Forest Preserve (MCFP), the TSC's priorities were to protect the existing ecosystems and unique endemic species in the area of the Continental Divide for biological research, educational purposes, and low-impact visitation by tourists. They determined that the area they acquired was too small to "preserve a representative sample of the biological and ecological complex of the Cordillera de Tilarán" (ibid.). Consequently, they continued to search for funding in the following years to expand the forest preserve, by inviting biologists and U.S. and Costa Rican environmental activists to visit the MCFP. In their international (mainly North American) fundraising efforts, they emphasized the existence of certain unique and charismatic species such as the golden toad, tapirs, and the resplendent quetzal; the forest's potential as a watershed; and the threat that agricultural expansion represented for the habitats (Powell 1974). They found financial support in diverse Northern sources, including the Explorer's Club, a German herpetological society, Philadelphia Conservationists, the New York Zoological Society, and private individuals in the U.S. (Powell 1989: 5; Tosi 1992: 2). By 1975, they successfully obtained an $80,000 donation from the World Wildlife Fund as part of the WWF's "Tropical Rain Forests Campaign" to assist land purchases and support administrative and patrolling costs for three years. Around this time, the owner of the 554-hectare Quaker watershed decided to sell his landholdings and move out of Monteverde. In response, Quakers and residents of Monteverde formed a stockholding corporation (Bosque Eterno, S.A.; or Eternal Forest) to purchase the watershed forest, and arrangements were made to include this land within the formally protected MCFP, to be administered by the TSC on a rental of one *colón* for ninety years.[10]

From the beginnings of the MCFP, supporters and organizers promoted the private ownership, management, and patrolling of the protected area, reflecting both their skepticism of the Costa Rican government's commitment to forest protection and the actual inabilities of the government to handle the preserve. During these years, the government's own environmental branch, the recently formed Servicio de Parques Nacionales (SPN; National Park Service), was struggling to consolidate itself bureaucratically (Wallace 1992; Gamez and Ugalde 1988; Boza 1993.). Costa Rican government decisions to establish and expand a system of national parks and protected areas have not necessarily publicized the same set of concerns about forests as in Monte Verde, though not necessarily for lack of common cause. According to Mario Boza, for example, he and other Costa Rican environmental activists working with the government in the 1970s aimed to protect areas of high recreational potential and cultural, historical, and scenic interest, "so that no one could object, making it easy to sell the public on the idea of conservation" (Boza 1993: 240). The SPN asserted that it would eventually seek the creation of a national park in the Monteverde-Tilarán mountain range area (an ongoing, though background, issue to this day), since that would provide the most secure form of tenure for the Arenal watershed.[11] But the TSC considered the political situation surrounding the preservation of lands in Costa Rica too volatile and uncertain to turn over the MCFP to government control. In particular, they were worried that pressures from lumbering and cattle interests and land colonization programs could threaten preservationist initiatives. This philosophy also sat well with the Quakers and other North American residents of Monteverde, who have been widely suspicious of Costa Rican governmental authority.

TSC and preserve administrators considered land purchases and patrols by guards to be direct methods of combating the settlement and further conversion of forest by land speculators and homesteaders near the boundaries of their growing preserve. According to TSC,

> In the selection of lands to buy, we have given preference to those that guaranteed a presence over both sides of the continental divide and the road to Peñas Blancas, where we could by our mere presence exercise an influence in braking the invasions into those national reserve lands inside. During this period, lands were bought along the road to "Chomogos" and in "El Valle," to eliminate land speculators in the zone and definitely close this trail as a public road (Tosi 1992: 3; my translation).

By purchasing lands at the continental divide, the TSC sought to gain control of the road that passed through Monteverde into the Peñas Blancas Valley. Despite the Valley's inclusion in the Reserva Forestal Arenal since 1977 (watershed reserve for the Arenal hydroelectric project), however, people who had land claims there sought to build a road to facilitate the movement of cattle, lumber, and eventually automobiles. In 1975, the Asociación de Desarrollo Integral de Santa Elena (ADISE, or Santa Elena Integral Development Association) obtained a *partida específica* (a government fund released for a specific purpose) to build a road into the Valley (Bulgarelli 1976). Road construction did not progress very far, however, before it was halted under pressure from the TSC and several government agencies (including the MOPT and the Instituto Costarricense de Electricidad) that opposed it on the grounds that it would lead to more deforestation of the Valley. These tensions would continue into the early 1980s as some landowners continued to press for the passage of a road through the MCFP.

Throughout the 1970s and 1980s, the MCFP gained increasing attention in U.S. and European environmentalist and scientific circles. The increasing visitation of scientists, amateur ornithologists, and other visitors, as well as a BBC documentary film crew that produced a 1978 film on the cloud forest of Monteverde

("Forest in the Clouds") that aired in Europe and the U.S., spread knowledge and images of Monte Verde cloud forests as pristine and unique wilderness. Attention had shifted away from establishing and bounding the preserve itself (approximately 5,000 hectares by 1987), and toward managing more guard patrols and visitors, especially as the latter numbers grew (around 6,500 visitors a year during the mid 1980s).

### *"You can see this forest from a satellite:" The Revival of Land Purchases*

By the early 1980s, the TSC had more or less finished expanding its preserve, dedicating its energies to consolidating control over its lands, and hosting increasing numbers of tourists. Concerned that the TSC did not appear to be committed to expanding its preserve, a number of North American residents and biologists in Monteverde village began meeting in 1985 to discuss the protection of Pacific slope forest habitat threatened by encroaching agricultural production. As a result of these meetings, they formed a new organization called the Monteverde Conservation League (MCL), which was legally constituted in April 1986 as a membership organization (hovering around 150 members during the past decade). This organization had astounding success in raising international funds for land purchases between 1987 and 1993, expanding the lands of the TSC's Monteverde Cloud Forest Preserve by approximately 5,000 hectares (to a total of 10,500 hectares), and creating the largest privately controlled forest preserve in Central America under direct MCL administration (currently over 22,000 hectares). Land purchases established the MCL's conservationist credentials with donors, and as a result it professionalized its staff and expanded into areas like reforestation, environmental education, and sustainable development schemes. In 1996, during a celebration of the organization's tenth year, one of its founders reflected, "I am happy and impressed with what we accomplished. You can see this forest from a satellite. As a biologist and conservationist, I feel we've attained something important here.... All these problems we've had, marches and conflicts, are natural in the life of an institution like this. Conservation goes against all of human history. I know it's a struggle, but we've done it with good intentions."

Interestingly, the initial catalyst that led to the MCL's creation was more humble, born of a highly local issue. As one of the MCL's founders explained to me, several dairy farmers in Monteverde village began discussing the development of a new water line from a nearby river for irrigation purposes. Other residents, artists and biologists among them, expressed concern that the black hoses running from the river could have negative ecological and aesthetic impacts on the landscape. A committee at the Monteverde Town Meeting was formed to look into the issue, and as this man told me, "these meetings were the spark that brought the biologists in the area together to talk about issues of conservation of natural resources in an organized forum, and really it's from there that the League was born, to protect Pacific slope forests."

That these people decided to establish a new organization, instead of working through the TSC, is significant in various respects. By the 1980s, there had been important changes in Monteverde village demographics and economic conditions, including a shift from a primarily dairy farming economy dominated by the Quakers to an increasingly touristic one in which the Quakers shared power with more recent immigrants, including biologists, retired North Americans, artists, and tourism entrepreneurs. In addition, tensions had developed between the San José-based TSC and residents of Monteverde, reflecting concerns about who should manage local natural resources. One Monteverde village resident explained to me,

> [The TSC's main interest seemed to be] in becoming an international consulting organization in the area of forestry management. People always felt they had invested little here. Many biologists thought that they didn't support research either. And if you weren't with them, you were outside, since they accepted very few new members. In this area, people tend to make their statements through institutions. So this is what the League was about, a local response to outside control over conservation.

From its earliest days, the MCL considered itself a self-consciously "grassroots" response to an organization they felt to be aloof and disengaged. For people who could not become members of the TSC, but were interested in taking actions to protect ecosystems and species, the MCL represented an important new channel for their activism.

In its early years, the MCL was considered by many Costa Rican residents of the Monte Verde region to be a North American initiative. Among its twelve legal founders, only two were native Costa Ricans. Early meetings were held in English, although it was quickly recognized that the MCL would never have a broader local appeal until it more actively integrated Spanish-speakers. Participants decided that their central goal would be the formal protection of forests, and the promotion of political action leading to the protection and regeneration of forests. They also decided that the most practical way to achieve these goals was to expand the formally protected area through land purchases. The majority believed that purchasing forests would be more effective than another option—leasing them from landowners—since in the long run it would be more cost-efficient for the organization. Like the TSC before them, they did not consider turning lands over to the government a desirable option, although not because the government did not have success since the early 1970s establishing parks. Rather, MCL participants believed there would be more local control over the lands if they were held in private hands.

Not long after they formed their organization, members declared a deforestation crisis in the Peñas Blancas Valley. The valley lies on the other (Atlantic) side of the continental divide from Monte Verde, and not on the Pacific slope that MCL members had declared to be their central interest. Through an arrangement with the TSC, volunteers were allowed to use the name of the Monteverde Cloud Forest Preserve in their efforts to raise money for land purchases in the valley, in exchange for turning whatever lands they bought over to TSC administration (CCT 1986). Still a volunteer organization, the MCL could not manage the lands itself, much less count on full-time fundraisers. But it could rely on dedicated and inspired advocates who gave lectures and slide shows and networked on behalf of the organization, among them North American and European biologists, residents, and regular visitors. Two well-known wildlife photographers shared photographs with fundraisers for use in the campaign. In town libraries, schools, and college campuses throughout the U.S. and Europe, people were introduced to the splendor of the Monte Verde region's landscapes and wildlife. Even a Grateful Dead concert was dedicated to raising funds to save the valley. Northern environmental media picked up the story early. For example, a Canadian biologist who worked in Monteverde published an essay on the MCL's efforts to preserve forests, in the Canadian magazine *Equinox* in 1986 (Forsyth 1986). According to one commentary, "In a few hundred words, he had presented 700,000 readers with a clearly defined project and a direct channel to participation" (Robinson 1988: 8). As a result, readers sent hundreds of thousands of dollars to the MCL through WWF-Canada to purchase rain forest. Canadian science celebrity David Suzuki also became involved in raising funds. In 1988, a WWF-Canada representative observed, "It was, without question, bar none, the most successful campaign in our history" (ibid.: 10).

One of the first large institutional grants that the MCL received for its Peñas Blancas campaign was from the World Wildlife Fund-U.S. It was a $25,000 matching grant that allowed the MCL to begin making purchases of forest and land claims in the valley, and even more importantly, provided it with credibility within international environmental conservation fundraising circles. Hundreds of thousands of dollars were channeled on the basis of this proposal, because international organizations like the WWF and The Nature Conservancy continued to use it to raise funds during the late 1980s and early 1990s.

Written by MCL volunteers, the grant proposal emphasized the remarkable diversity of species found in the unique cloud forest ecosystems of the Monteverde Cloud Forest Preserve. It also emphasized the urgency of the threat to the valley lying below the preserve, reflecting powerful and morally loaded language regarding Costa Ricans and their destructive relationships with forests:

Land owners in the Peñas Blancas valley are willing, even anxious to sell out. The area has proved unsuitable even for subsistence farming because of heavy rainfall, poor and unstable soils, steep terrain, and lack of road access. Although the Tropical Science Center has been negotiating land purchases with Peñas Blancas land holders for eight years, they have been unable to raise enough funding. As a consequence, the owners have become frustrated and several of them are acting to pressure either TSC [Tropical Science Center] or the government to act. This pressure includes encouraging squatters, forest clearing, and lobbying local authorities to put in a road that would open up the area to rapid lumbering and colonization.... *If this movement is not stopped immediately or if pressure mounts to put through a road, the entire watershed may be lost and the present Reserve jeopardized. THIS YEAR IS IT!* ... *If the MCL is unable to accommodate these people, their only option will be to push for the destruction of the Peñas Blancas habitat* (MCL 1986: 3; emphasis added).

The language of a crisis in the valley is an important innovation here, one that the TSC did not employ as effectively in its earlier efforts. The imagery of inevitable destruction at the hands of *campesinos* suggests that concerns identified as scientific—ecosystem and species survival, for example—are not necessarily enough to provoke action. The imagery of the destructive settler or speculator is a culturally loaded one, even though by the proposal's own admission, only two or three individuals (out of sixty landowners in the valley) were responsible for most of the pressure to convert forest. In fact, in over forty years of settlement and speculation, the area the MCL proposed to buy was 92 percent "pristine virgin forest" (the proposal's description) reflecting perhaps the *non-inevitability* of widespread forest destruction at the hands of local settlers.

It would be inaccurate to assert that rural Costa Ricans were innocent of forest conversion in the valley. But it is important to ask why several landowners threatened to cut or did cut trees while most others did not, and why imagery of poverty-stricken and avaricious peasants does not apply in this case. Part of an answer lies in the fact that most landowners in the valley (among them Quakers) were not subsistence farmers who needed this land to survive (only nine actually lived on their lands), but considered their landholdings to be long-term investments or alternative cattle pasturing for times when the Pacific slope of Monte Verde was in dry season or drought. Aside from the dairy industry the Quakers had established, there were no strong economic forces driving the conversion of the forests. Furthermore, even during the 1980s, landowners' intentions were not necessarily the immediate and inflexible destruction of the forests, for as soon as negotiations over land purchase began, cutting stopped. Landowners and speculators considered the threat to cut forest as a symbolic political action to provoke either government or environmental groups to force a resolution to their problem of frozen lands stemming from the 1977 declaration of the area for hydrological protection.

And yet the image of inevitable and immediate destruction is both compelling and persistent, reproduced throughout the history of Monte Verde land purchase campaigns. International media accounts, which increased as big international organizations like WWF and The Nature Conservancy brought attention to Monte Verde, emphasized the imagery of land hungry farmers and even employed exaggerated figures to demonstrate the crisis of forest destruction. For example, an article on Monte Verde in a popular Canadian weekly opens with an image of "poverty-stricken migrant farmers [who] have cleared ['once lush and verdant woodland'] to provide land and fuel" (Dwyer 1988); an image that contrasts sharply with the actual situation of non-poverty-stricken migrant farmers, or the fact that few used such wood as fuel. The same article mistakenly claims that "Costa Rican squatters have destroyed more than 50 percent of the country's rain forest within the past 15 years" (ibid.). There is also substantial conflation of Monte Verde's history of forest loss with the contemporary loss of tropical forest in other parts of the world. For example, in a fact sheet sent to prospective donors from the late 1980s, the MCL made no specific references to threats or rates of forest conversion in the Peñas Blancas Valley or the Monte Verde region itself. Reference was made, however, to information such as "Tropical forest destruction proceeds at the rate of 50 acres per minute. At present rates, 20 percent of the world's remaining tropical forests will be destroyed by the year 2000" (MCL, n.d.). Similarly, a Nature Conservancy "Conservation Abstract" that was sent to donors specifically to raise money for Monte Verde asserts that "At present rates, all unprotected primary forest [in Costa Rica] will be gone in less than ten years" (TNC n.d.). The assumption (perhaps accurate) is that donors will not care so much about Monte Verde as they will about the crisis facing tropical forests in general.

In considering which landowners to approach and lands to purchase in the valley itself, MCL land-purchase strategies and criteria varied, depending on institutional priorities, opportunism, donor intentions, clarity of land-ownership records, and the desire to control access and use of the forests. Aside from some interest in conserving habitat of the migratory quetzal (a factor that was important because of conservation biologists influential in the MCL), scientific and technical criteria tended to be minimal in the expansion of the growing preserve. The contingencies of negotiating land purchases were sometimes related to the opportunism of landowners and speculators themselves, some of whom tried to sell land to which they had no legal claim, or who argued their land claims overlapped with the claims of others. To some extent, this stemmed from the ambiguities inherent in rural Costa Rican traditions of land settlement and ownership, especially the fact that people did not necessarily hold title to their lands, reckoning their ownership through eminent domain, as the *mejoras* they made, or through informally accepted boundary agreements with neighbors.

In its first Peñas Blancas land purchase campaign, which concluded in 1989, the MCL bought some 6,200 hectares at an average price of $35 per acre. More importantly, the campaign was judged successful by the amount of money it generated from U.S. and Canadian sources (almost a half million dollars). It motivated the World Wildlife Fund to invite the MCL to participate as one of the beneficiaries of Costa Rica's first "debt-for-nature" swaps, which made hundreds of thousands of dollars available to continue land purchases, and embarked the MCL on other fundraising campaigns. It encouraged a massive expansion of the MCL itself, moving it toward a professional staff and into areas like environmental education and reforestation. But it also laid the groundwork for growing conflict between the MCL and TSC, which has only recently been resolved. The agreement they had reached in allowing the MCL to use the name of the "Monteverde Cloud Forest Preserve" in the Peñas Blancas campaign was never formalized, although the TSC became the *de facto* administrator of the lands bought under this arrangement. The TSC claimed that since the MCL had used the name of the MCFP to raise funds, it was the rightful owner of the lands. Throughout the 1990s, the unresolved debate generated tensions between the organizations, and the TSC took the MCL to court over the issue in 1996. Several years later, they settled out of court, the TSC ceding those lands to the MCL. Although the size of the MCFP had once been 10,500 hectares, it was reduced to its current size of 5,000 hectares, and the MCL's preserve expanded to its current size of 22,000 hectares.

Beyond this conflict, the Peñas Blancas Campaign generated certain ironies, mostly out of MCL control. One of these is that the MCL, a group self-consciously established to change how people value tropical forests, paid more for "disturbed habitat" than forests. Indeed, until the Forestry Law of 1996, rural Costa Rican practice has been that the labor that went into cutting a forest sets the value of a land claim, not the forest (up to a 50 percent difference). Or that within several years, after much success purchasing lands, the MCL was priced out of continued land purchases as landowners and speculators pushed their prices higher, knowing the MCL's internationally funded deep pockets. Within several years, land values inflated from around $75 per hectare to an average cost of between $160 and $350 per hectare, and by the late 1990s, upwards of $1,000 per hectare in certain areas.

But during their land purchase campaigns MCL land negotiators tended to be most concerned with moving people out of the areas considered desirable and important to protect, specifically primary forest, in areas that they knew they could afford. They were not particularly interested in redefining the values placed on the differential valuation of deforested land and forest, or working with the owners of those forests to find alternative ways to remain there (with one significant exception, as explained in chapter 5). As one MCL land purchase agent pointed out, "there will always be conflict and overlapping, but we were willing to pay people twice for an overlap. The point was to close the deal and get the people out of there so they don't do any more damage." That is, landowners, who were conceived as "obstacles" inherently hostile toward land purchases, if not tropical nature itself, needed to be removed. There were ways to get recalcitrant landowners to sell by suggesting that their access to their own lands would be restricted.

In spite of such tactics, which were a source of frustration to some landowners, landowners lined up at the hotel where land purchases were being coordinated to sell their lands. MCL founders and officials largely considered that they were performing a favor for landowners by offering to buy land claims in areas where government decree had prohibited the development of farms or timber extraction. They were also sure that landowners would find economic opportunities elsewhere, and MCL activists expressed little, if any, interest in the potential social consequences of the land purchases. There was, at least for some MCL activists, an assumption that these landowners had lived a spurious agricultural tradition, or as one volunteer working in land purchases explained to me, "I've always felt that people here could change easily, since this was a new agricultural tradition."[12] The idea that people could live in a wilderness area in the first place, and not inevitably destroy it, remained unquestioned.

And yet the scale of the Peñas Campaign paled in comparison with what would come next, as a transnational movement of children "saving the rain forest" emerged and chose the MCL and Monte Verde as its first major site to save.

## Children "Save the Rain Forest"

It is worth pausing here to note that there is nothing inevitable about the processes that led to the creation, expansion, and legal consolidation of nature preserves in Monte Verde, either in the TSC's efforts during the 1970s or in the MCL's efforts during the 1980s and 1990s. This observation holds true even if we admit the vaguely scientific and intellectual grounds upon which both to dismiss rural Costa Ricans as destructive and to judge Monte Verde's endowment of tropical nature to be outstanding and worthy of formal protection. These latter concerns explain neither why large areas in the Peñas Blancas had remained forested, nor why Monte Verde, which is only one site where cloud forests exist in Costa Rica, became an intense focus for formal protection efforts during the late 1980s and early 1990s.

In fact, the success of environmentalist land purchase campaigns were made possible by certain political-economic processes and inequalities in regional markets and development, which coincided and articulated with environmentalist fundraising efforts. There is little doubt, as I discussed in chapter 2, that this dramatic consolidation of land was facilitated by the government's declaration of certain areas like the Peñas Blancas Valley as protected hydrological reserve, and that certain farmers were experiencing diminishing returns from their lands, so they were willing to sell. It also relied on the increasing visibility of Costa Rica as a peaceful and tourist-friendly country in a war-torn region, as well as the fact that the U.S. and European-based environmental media had picked up the Costa Rica (and Monte Verde) story.

Recognizing that children and their donations constituted a growing audience for its message and source for its fundraising, in 1988 the MCL renamed its expanding forest preserve the "Bosque Eterno de los Niños" or Children's International Eternal Rainforest (BEN) and embarked on a new fundraising campaign. This campaign had even more astonishing results than the previous Peñas Blancas efforts, expanding the area the MCL controlled to some seventeen-thousand hectares by 1993. Its preserve now borders on communities with very different histories and relationships to nature conservation than Monteverde village and Peñas Blancas Valley, stretching its administrative capabilities to new levels, and generating significant dilemmas in how to manage such a large area.

The origins of the BEN fundraising campaign are in Sweden, and reflect a "grassroots" image proudly promoted by the MCL. In 1987, a North American biologist (the woman whose passionate reflections on riding the bus from Monte Verde opened this chapter) was invited during a sabbatical leave to work in Sweden. During this trip, she made a visit to an elementary school in Fagervik where she gave a slide show on the Monte Verde cloud forests. A boy in the group suggested that he and his classmates could raise funds to support efforts to save rain and cloud forests. The teacher, who coordinated the resulting efforts to raise funds, explained that the initiative was a response to powerful media messages the children had experienced,

> The favorite television programs of these children were films about nature that always ended: "We do not know how long this animal will survive on this planet because its habitat is disappearing." The kids wanted to do something about it. They saw a television program on the work of Daniel Janzen in Costa Rica and decided to collect money for the purchase of rain forest. They were convinced that I would find a project and tell them "Yes, let's begin and maybe others will follow." They began to make drawings and books to sell. ... A few weeks later [the biologist] visited the "rain forest" children in the Fagervik school and showed the beauties of Monteverde in slides (MCL 1991: 6; my translation).

Out of this visit, Barnens Regnskog [Children's Rain Forest] was born as a nonprofit organization set up to raise and channel money to the MCL for land purchases. Children from all over Sweden sold paintings and Christmas cards, gave pony rides, told stories about rain forests, volunteered work days, wrote and performed songs and skits, and asked for and gave acres of rain forest for Christmas (MCL 1988: 4). Between 1988 and 1992, Barnens Regnskog raised $2 million for land purchases from over one hundred thousand people (Burlingame ibid.). In addition, they arranged for SIDA, the Swedish overseas development agency, to donate $80,000 to support MCL reforestation and guarding efforts (an amount that ultimately grew to $350,000) (Guevara 1990). In the U.S., England, Germany, Austria, Japan, and even (though to a much lesser extent) Costa Rica, similar organizations were established by adults and children interested in "saving" Monte Verde, many of which were also organized around school classrooms. The national organizations formed the "International Children's Rain Forest Network" to coordinate fundraising efforts, promote cultural exchanges between members, and review projects in other parts of the world to finance. By 1990 these member organizations had sent over $2 million to the MCL, most of it for land purchases (Zuñiga 1990).

Participants in the Children's International Rainforest campaign have attributed their success at raising funds to the proliferation of images, films, and magazine articles on the destruction of the world's rain forests, stories of the biological riches of Monte Verde forests, and the television and newspaper reports of children raising funds to preserve tracts of forest in Monte Verde and elsewhere. But, as activists have pointed out, "No one planned on the children. But they—with energy, creativity, considerable skill, and intrepid morality—became the leaders and the reason for the project" (Kinsman 1991: 9). Moreover, the focus on land purchases to preserve distinctive habitats in which charismatic species live (especially the golden toad and resplendent quetzal) represented "an aspect of conservation most understandable to and rewarding to children" (Children's Rainforest U.S. 1994). Children in these fund raising efforts embraced ownership of Monte Verde flora and fauna: "They talk about *my* forest, *my* resplendent quetzals, *my* golden toads. They have a sense of pride and responsibility" (Austin 1990: 47).

For an elementary school teacher from the U.S. Midwest who with her students raised over $20,000 (and who contributed $20,000 of her own money), having children commit themselves to "save the rain forests" of Monte Verde reflects a lesson in taking active leadership of global problems. During a trip she made to Monte Verde in the late 1990s, she explained to me her reason for promoting the involvement of her classrooms in raising funds: "If you are going to tell kids about tropical deforestation, you have to give them a tree to plant in return. You have to be part of the solution, and I knew right away when I heard about it that raising funds for BEN would be a wonderful way for my second graders to learn about the rain forests. It means a lot more if they're actively involved." In this case, "planting a tree" is an interesting, if inaccurate, metaphor, in that adopting acres had nothing to do with regenerating already-degraded forests (a concern conservation biologists had already begun expressing in Monte

Verde), but sending money abroad through certain intermediaries to "lock in" the already-pristine forest. This practice is based on a well-intentioned idea that the problems of suburban U.S. children are the same as those of rural Costa Ricans. Dahl (1999: 26) describes such situations somewhat critically as the difficulty for citizens in one place to understand the conditions, situations, needs, wants, aims, and ends of other citizens who are distant and different from themselves. Although the highly depoliticizing image of children here has largely blunted such critiques in Monte Verde.

But the success of this campaign was not without dilemmas for the MCL. Reflecting institutional policy, officials assured donors that they would spend their funds however the donor wished. Therefore, if a donor desired to "adopt-an-acre," employees would direct the money to the land purchase accounts. MCL officials could claim their preservation mission to be morally upright, because not only were they honest to donors' desires by purchasing land, they had the direct support of children from many countries to do it. At the same time, however, they found it difficult to redirect funds to more pressing issues. So in the early 1990s, recognizing that their expanding preserve required greater financial resources to protect and manage, MCL officials began publicizing a new fundraising effort called "Rain Forest Partners." This international fundraising campaign, directed at the same people who had helped adopt acres before, solicited donations for pay for forest guards, infrastructure, and institutional development, instead of simply land purchase, "to broaden the conception of what protection of rain forest actually encompasses for the MCL" (MCL 1992a). The lackluster response of donors to the campaign suggests that the fad of adopting acres represented a more compelling and tangible purpose, not to mention that international donors were not always obedient to or interested in local concerns.

By 1993, MCL administrators discontinued land purchases, citing administrative concerns, including the difficulty of patrolling its now very large territory. This cessation was in spite of agitation among some of its members to continue expanding protected areas, because, as one biologist told me, "the formally protected areas didn't necessarily conserve what's representative in this area." But administrators were concerned that as their protected area grew, donations for programs such as environmental education and protection did not match the amounts necessary to expand those programs into new communities (MCL 1994a: 7–8). Importantly, funding priorities also began to shift within some of these international organizations, so that Costa Rica, much less Monte Verde, became less important sites of conservation investment. For example, although the MCL has maintained an important relationship with the Children's International Rainforest into the present, members of the network began funding the creation of "International Children's Rainforests" in other countries in the mid 1990s, reflecting their desire to achieve the successes of Monte Verde elsewhere. Ironically, this very success has undermined the MCL's own fundraising base. As a North American man representing an international fraternal order explained to me in 1996, his organization would not send any more donations directly to the MCL or other Monte Verde environmental organizations: "We aren't interested in giving here anymore, because they've been successful. We're trying to find new places to give money to." Combined with a general decline in Northern concern in Southern rain forests since the late 1990s, the MCL has been in difficult financial straits, and other initiatives (such as a bellbird conservation initiative born in the late 1990s) have confronted severe difficulties trying to raise international funding.

The challenge presented by this drop in international donations is formidable. From 1997 until the present (2005), the MCL has been more or less in "survival mode": most of its programs (environmental education, reforestation, biological corridors) were terminated by the mid to late 1990s, and since they had never dedicated any funds toward an endowment, administrators' main concerns revolved around searching for ways to secure the financial future of the organization. One option that has been frustratingly slow in bearing fruit is ecotourism, since the MCL had constructed two biological stations in the BEN (San Gerardo and Poco Sol) during better financial times. At the same time, MCL administrators were aggressively lobbying at the national level (with success) to have the tax burden on their vast landholdings reduced. They also sought to secure funds from "Joint Implementation," an international funding scheme motivated by growing concerns over climate change and global warming that raised money in the industrial North to support "carbon sinks" in the global South. One source of revenue came in 1998, when the MCL signed an agreement with a private hydroelectric company to receive up to $30,000 per year for ninety-nine years for the watershed protection provided by the BEN (Burlingame ibid.: 367). In 2005, a U.S. fundraiser for the MCL created a $1.5 million "Land Purchase and Protection 20th Anniversary Campaign," reflecting an agitation for land purchases that has not disappeared. But the MCL, while still wealthy in lands, is a shadow of the powerful organization it once was just a decade before, with half the personnel and an administration in relative disarray.

## Part Three: Land Purchase Campaigns and Local Democracy

Land purchases have been both the most tangibly successful and the single most controversial environmentalist practice, sparking the most public debate and rancor (Vivanco 2002b). Conflicts over land deals and the actions of forest guards enforcing antihunting laws have led to deep grudges, as well as specific acts of revenge and protest, including one situation in which an ocelot was killed to protest what some members of one family considered an unfair interaction with land purchase agents, as well as acts of vandalism against the properties of both the MCL and the MCFP, including an attack on the MCL's tree nursery. There have also been personal attacks on administrators of both organizations. While some of these tensions were personal in nature, there were also collective acts of protest against land purchases. Even while landowners lined up at the MCL office to sell their lands, there was considerable concern throughout the region that the MCL and its international funders were aggressively displacing landowners. In 1990, those tensions came to a head when the local Catholic priest, Santa Elena high school teachers and students, and several Costa Rican entrepreneurs organized a march on the MCL office. One participant described the march as a response to rumors that the "land-hungry" organization was secretly lobbying to buy a government property under the control of the high school (the Reserva Santa Elena). MCL officials, a few of whom met the demonstrators in front of their offices, publically denied such rumors. But the appearance of impropriety was already widespread, reflecting a concern that land purchases threatened Costa Rican control over natural resources in the long term. Residents felt their resistance justified when they learned that land ownership was being temporarily registered under the names of individual employees, because of land ownership laws that prevented the concentration of lands beyond a certain quantity by legal associations like the MCL.

By the mid 1990s and into the present, land purchases became difficult because of higher land prices and a lack of funding. There was also a shift in attitude at the MCL, where previously avid promoters of land purchases were cautious about reinitiating new campaigns. For example, one former MCL land purchase agent told me that although he once worked hard to expand the MCL's preserve, he no longer supported land purchases since they do not reflect the social realities of communities where the last formally unprotected forests exist:

> I do not think it is possible any longer to buy lands for conservation, and I don't agree with efforts to do it. The last big forests are in much more populated areas, so it would be difficult to ask the people who live there to leave. It would also take incredible amounts of money. We need to find alternatives such as conservation easements, that would keep farmers from cutting their forests, instead of doing whatever we can to get them out.

The idea of conservation easements emerged in the 1970s in the U.S. as a mechanism to ensure the preservation of historical buildings and landscapes while preserving private ownership of them, by promoting the joint creation of contracts among neighbors that specify mutually agreed upon limitations on use. In both the MCL and the Monteverde Institute (MVI), these discussions have gained increasing importance in recent years. For example, the MVI has an initiative called Enlace Verde ("Green Links") that since 1994 has promoted the conservation of forest on private lands in Monteverde village and the establishment of conservation easements between neighbors as a way of promoting both forest preservation and land-use and community planning.[13] One potential advantage of easements is that the tradition already exists in rural Costa Rica for rights-of-way and other negotiated easements. Nevertheless, outside Monteverde village, such arrangements are still quite uncommon.[14]

In spite of such alternatives, many people have remained skeptical not just toward the MCL, but also toward other environmental organizations and their initiatives, with the result that some new initiatives have been blocked. For example, in the mid 1990s, the TSC was greeted with profound skepticism and rejection when it proposed a biological corridor through San Luis, a village in the Monte Verde region that borders the MCFP (CCT 1995). Some of the residents had once lived in San Gerardo, an area that fell within the MCL's expanding Children's Eternal Rainforest, and argued to neighbors that the organization's ulterior motive was to control their lands, as the MCL did in San Gerardo. Commenting on the TSC plan, one former San Gerardo landowner argued: "Nobody should come from the outside anymore and tell us how to manage our community. We can guide it ourselves. They shouldn't impose on us. They seem to love nature more than us. What they don't understand is that we love nature too." Whether it was true or not is more or less irrelevant, but as a result of such skepticism, the TSC's efforts were significantly complicated and indefinitely delayed, because it could not prove to its donors that it had secured community support and participation.

Underlying such concerns is a conviction that land purchase campaigns have tended to consider landowners and farmers to be obstacles to nature protection, and it is perhaps not surprising that some in the Monte Verde region have considered land purchase campaigns threats to Costa Rican sovereignty. Tapping into that widespread frustration, a study of conservation land purchases in Monte Verde by a team of Costa Rican scholars concluded that land purchase processes were "extremist" (Vieto and Valverde 1996), and various *campesinos* told me they considered the land purchase campaigns to be "antidemocratic." But it is important to place such claims in perspective. When Monte Verdeans have referred to environmentalist land purchases as antidemocratic, they were not necessarily referring to the Costa Rican system of formal representative and institutional democracy. Rural Costa Ricans may in the abstract be quite proud of their country's tradition of formal electoral democracy, especially when compared to neighbors like Nicaragua or Panama (Edelman and Kenen, ibid.; Wilson 1998). But for people whose land becomes a formally protected hydroelectric reserve without their knowledge, and then who see more politically connected and wealthy landowners gain favorable treatment through expropriation (as happened to many San Gerardo residents when the Arenal watershed was declared as a protected area), the formal institutions of Costa Rican democracy may exist to serve the privileged rather than themselves.

In fact, cultural meanings and expectations of "democracy" and "justice" have daily realities that may be quite distant from processes and institutions of centralized governance, reflecting vernacular meanings and connotations through "common sense" (Lummis 1996). In this respect, polemical accusations of environmentalism and environmentalists as "antidemocratic" have less to do with formal institutions of national governance and abstract rights, and more to do with the processes of social life and conventions of conviviality that characterize rural life, and how environmental activists and organizations did or did not fit those specific practices of social custom.

Most Monte Verdeans recognize that environmentalism is a movement with highly political goals and methods, whose intention is no less than to reorganize how people in the countryside will relate to the natural landscape. They have also understood that powerful—and wealthier—international actors and interests funded land purchase campaigns. What these various actors—certain MCL administrators, donors, and the like—have not necessarily appreciated or accepted, though, is that some of the rural Costa Ricans with whom they have dealt maintain distinctive ideas about political processes and public interactions, ideas that are embedded in local social relations. One Costa Rican who worked in the MCL as a reforestation officer explained that land purchases sparked significant controversy precisely because of these differing concepts of negotiability that he and other Costa Ricans take for granted:

> It is becoming more clear that these mountains no longer belong to the national community. They belong to the environmental organizations, their biologists, their foreign funders, students, the tourists, the tourism entrepreneurs. Because there was no negotiation, a space in which to search for agreement, we could find ourselves living a situation that is neither magic nor imaginary, but concrete and real. I do not mean to be so apocalyptic, but sadly, all this hard work to protect forests could be undermined by people who do not care about these forests and are willing to invade them, simply because conservation did not care to establish a communication and a sharing of information, participation, and shared interests. And this is not because people are ignorant or do not know the value of forests. Culturally, we are close to the concept of the pact. What is a pact? It is an assembly of dissimilar interests, with diverse necessities, but that are disposed to meet on the grounds they share, and together find ways to further their common interests. This is also not magic. Negotiation, a pact, communication are not a gift of divine providence. It is something to construct with intention or a true practical sense of opening, of learning to listen.

Central to this polemic is the perception that environmental groups and their international funders have been primarily concerned with outcomes, such as formally protecting a certain number of hectares of rain forest or building infrastructure to attract tourists, and they never came to recognize the conflicting political, social, and economic interests of rural people. The tendency throughout land purchase campaigns, in fact, had been to often reject the claims and interests of rural people and to delegitimate their specific claims and histories. This tendency has posed a fundamental challenge to local concepts of democracy as being based on values of recognizing and airing differences and of openess to negotiation, thereby affirming common bonds instead of undermining them. Furthermore, this commentator raised a provocative challenge for environmentalism, asserting that rural Costa Ricans have been thought of primarily as passive subjects to be acted upon, only collaborated with insofar as it helps achieve certain institutional, strategic, or ideological goals. In this sense, the land purchase campaigns have ensured the centralized control of large territories, generating a concentration of land in an area that historically was composed of small and medium-sized landowners. The point is that if processes are not based on an attitude of openness and willingness to listen to diverse visions of the people who also have a long-term stake in the landscape, the grounds for a richer, more inclusive and ultimately stronger environmentalism remain elusive. It also suggests why the MCL's reforestation projects have had greater popular success than have land purchases, because they enrolled landowners as participants, not obstacles, and provided concrete benefits for them in terms of enhanced agricultural production in a working landscape.

There is no doubt that this is a strong and polemical indictment of land purchase campaigns, and many (especially proponents of land purchases) would vigorously disagree. However, there is also no doubt that land purchases have had a significant effect on local concepts and practices of property ownership: specifically, the reification of property boundaries and clarification of title in a region where concepts of ownership and rights of passage have been relatively fluid and flexible. Reflecting on this, one Costa Rican man described a walk he took with his young son. When he suggested to his son that they take a shortcut through a forest to get home more quickly, the boy's reaction was that forests belonged to environmentalists and were off-limits. Interpreting his story, the man explained that this is a negative reflection of the absolutism of environmentalist ideas about protecting the land, in an area where conventions of property ownership were never absolutist, and where neighbors certainly could walk through each other's lands. Under the surface is a concern that people who once had a stake in the landscape no longer do, or as another man who was present at the telling explained, "Growing up here we were always told that these forests were our patrimony. But now they are not." Central to this are new forms of control over peoples' movements throughout the now-protected landscape and their use of natural resources. Because those decisions are made in the context of centralized institutions who patrol the landscape like the MCL and TSC (not in the flow of everyday interactions), and with an absolutism backed by legal authority and conviction, some long-term residents have expressed a sense that the concentration of lands as preserves has to some extent undermined local conviviality and informal democracy.

# Notes

1. Peter Raven is director of the Missouri Botanical Garden and a prominent figure in international environmentalism. This lecture would have emphasized the causal relationship between human population, consumption, and environmental degradation, and the urgency of tropical conservation (cf. Raven 1986).

2. OTS (Organization for Tropical Studies) is a U.S.-based consortium of universities that organizes courses for graduate and undergraduate students in tropical ecology. Between the early 1970s and early 1990s, OTS brought student groups to Monteverde to study cloud forest ecosystems. It stopped bringing groups to Monteverde because it became too crowded with tourism. In 1996, they resumed bringing courses to the Monte Verde region.

3. This trend seems to have slowed, as land prices in Monteverde and the Monte Verde region have risen, and fewer biologists have come to Monteverde to do research.

4. There are several possible exceptions to this. The Monteverde Cloud Forest Preserve has supported researchers with basic facilities that researchers themselves have had to equip. The Monteverde Institute, founded in 1986 to bring North American ecology study groups to Monte Verde to fund "community development" initiatives, is another possible exception. They have archives of some Monte Verde research, although researchers complain that this is still not enough. As a result, several biologists who reside in Monteverde have served as informal repositories for most ecology information. Another institution that attempts to maintain updated information on regional ecology is the Monteverde Biological Station, a private residential research station that hosts North American, Costa Rican, and European ecology student groups, and an important site where ongoing research is presented in open fora when study groups are in residence.

5. Monteverde is sometimes compared (unfairly and inaccurately) to the biological research community based at La Selva Biological Station, a research station in the Atlantic lowlands of Costa Rica operated by the U.S.-based academic consortium Organization for Tropical Studies (OTS) (McDade and Hartshorn 1994). Monteverde does not have the research facilities or the administrative centralization of La Selva, and is an informal nonpurposeful community where people also live, as opposed to mainly being there to carry out biological research.

6. For these reasons, ecological theorizing about the effects of nature reserves and landscape fragmentation on the evolutionary viability of populations have drawn on island biogeography theory (MacArthur and Wilson 1967). Island biogeography theory has offered a way to think about the relationship between species diversity and stability, as well as the role of disturbances (i.e., invasions, tree falls, etc.) in contributing to species extinction. It is based on an analogy between an unpeopled island and a nature preserve, and presumes to offer criteria to determine the "minimum critical size" that a preserve must be to maintain stable populations.

7. The Tropical Science Center was founded in 1962 as a forestry consulting organization. Its membership consists of approximately fifty prominent Costa Rican and North American scientists, forestry engineers, and academics.

8. In 1969, the Costa Rican Institute of Electricity (ICE), preparing to construct a hydroelectric dam in the Arenal region north of Monte Verde, took its first steps toward ensuring the watershed. It established a limited three thousand-hectare "electric energy reserve," in order to "guarantee that the resources of this watershed are transformed into electric energy useful for the development of the country" (ICE 1978: 3). During 1977, in coordination with the Ministry of Agriculture and Cattle (MAG) and decreed by law, a much larger Arenal Hydroelectric Forest Reserve was declared. In these reserves, cutting trees, burning, hunting, and agriculture were prohibited without the permission of the Dirección General Forestal.

9. Since the 1940s *campesinos* from the Monte Verde region, and later several of the Quakers, had been entering the Peñas Blancas Valley east of Monteverde on the Atlantic side of the Conti-

nental Divide, to establish claims by clearing forest and growing cattle pastures. Aside from this agrarian settlement and timber extraction, during the 1950s a Canadian mining company had begun to extract sulfur in the lower Peñas Blancas Valley.

10. To some extent, this represents the formalization of a broader Quaker interest in the preservation of the watershed. When the graduate student left Monteverde in the mid 1970s (to return periodically since then as a part-time resident and researcher), he left behind important converts to forest preservation, including one of the Quaker men who worked as a biological field assistant and as a forest guard at the Preserve. The contract between Bosque Eterno, S.A. and the CCT comes up for renewal every five years. In 1995, after an intense five-year growth period in which visitation rates skyrocketed, some stockholders strongly questioned the CCT's intentions with the MCFP, wondering if they viewed it as a moneymaking operation.

11. This discussion—to establish a national park in the Monte Verde/Tilarán mountains area—has been ongoing since the early 1970s between government officials and representatives of the CCT and the Monteverde Cloud Forest Preserve (and since the mid 1980s the Monteverde Conservation League). This discussion has been spurred by the importance of this watershed not only to the Arenal dam project, but to a major irrigation project in the Tempisque lowlands of Guanacaste.

12. This idea runs counter to the other main indictment of *campesinos,* which is their inability to be flexible because of a blind attachment to traditions. With their focus primarily on educating children, environmental education initiatives in Monte Verde have been based partly on this philosophy that older people cannot change, and therefore efforts to educate people should be directed toward "fresher minds."

13. In 1996, the Monteverde Institute received a grant from CEDARENA, a San José-based NGO promoting easements, to pursue easement work in Monte Verde (Scrimshaw et al. 2000: 382).

14. For example, in the late 1990s, Enlace Verde was working with over fifty properties, although by 1998 only three were written into deeds (Scrimshaw et al. 2000: 382).

# II Part Two: Landscapes and Lives: Environmentalism's "Social Work"

"… We have begun to realize that conservation is first of all a human social issue. Conservation activities are primarily designed to modify human behaviors that affect biodiversity."

(Salafsky 2001: 185)

In both theory and practice, the conservation mainstream tends to perceive its work as divided into natural and sociocultural sides, reflecting the separation typically made between people and nature "out there." For Monte Verde environmental organizations, purchasing land initially represented a key focus as a way to forge a physical separation between ecosystems and the people identified as threatening to those ecosystems. But conservation is never simply about what kind of nature people imagine or know they want to preserve or restore; it is also an important arena in which they, explicitly and implicitly, project and reimagine social relationships, cultural attitudes, and political institutions. Indeed, the two should be considered as co-constructed processes: through conservation itself, new and authoritative visions of nature are derived and implemented, just as culture and social relations are scrutinized, redefined, and normalized (Vivanco 2003).

In fact, purchasing land as a direct means of "saving nature" has not been the sole and explicit means of action for environmental institutions in Monte Verde. As I mentioned in the last chapter, some individuals actively contested the emphasis on land purchases within their own organizations, especially the MCL, pushing successfully for engagement with the region's people and communities through educational programs, reforestation initiatives, and other participatory projects. One of these initiatives, the MCL's reforestation program, which lasted between 1989 and 1994, has been considered by many residents and environmental activists to be unequivocably successful. But the director of one of Monte Verde's dominant conservation organizations put it to me this way: "We have to balance our efforts. We can think of ourselves as *latifundistas* [big landowners], and focus our energies internally on operating 'the farm,' such as protecting it from hunters, finding out what is inside it, bringing visitors to it, and so on. Or we can be externally focused, working with the people who live on the peripheries of our lands, in hopes that our work with them will convince them to support our efforts to conserve." Although this stark "either-or" scenario is overstated, the comment refers to one of the most profoundly and consistently contested themes among Monte Verde environmental activists, that is, how to design, communicate, and implement conservation initiatives in ways that involve local people and their community institutions in, as the epigraph expresses above, modifying "the human behaviors that affect biodiversity."

It is tempting to attribute this concern to the abstractions of "sustainable development," given that peoples' livelihoods and participation are central tenets in the retooling of modernist development to be more "green" and nature conservation to be more people-centered (Adams 1990). It is most famously expressed in the conviction that economic development is self-defeating in the long run if it does not contribute to the conservation of natural resources; but that nature conservation is bound to suffer, if not fail, if it does not address local peoples' livelihood concerns, directly involve them in the efforts to conserve natural resources, and create alternatives to resource-intensive productive activities (Wapner 1996: 83). Another way to express this is that the natural landscape can no longer be bounded in any unproblematic way, and therefore its conservation must address the character of social relations surrounding and intersecting with it.

As I will explore in Part Two, the rhetoric of sustainable development gains meaning and relevance not simply from its universalizing worldview and ambitions, because the specific social settings and encounters in which such principles are applied are typically characterized by diverse and unsettled meanings surrounding its most basic categories, such as "community" and "participation." The purpose of this part of the book is to examine some of the complexities, contours, and tensions generated when the targets of environmental initiatives express their obedience to, and rejection of, efforts to involve them. It does this by illustrating some of the forms of popular and communal participation imagined and permitted by environmental activists and organizations in Monte Verde. It also explores how efforts to generate participation become arenas in which culturally nuanced debates over appropriate forms of social action take place, and themselves can contribute to the creation of important new institutions, cultural meanings and identities, and political and economic possibilities for people living in Monte Verde. Central to these concerns are the following key questions: Who can be an environmentalist, and what does it mean to be one? Who is really in control of defining and carrying out environmental initiatives? If it is agreed that involving people though participation and community programming is crucial for the protection of natural resources, should not the boundaries, norms, and goals of environmentalism be potentially modified, if not altogether redefined, to accommodate diverse perspectives, local sociocultural realities, and the needs of targeted people?

The answers to these questions do not exist in a cultural or historical vacuum, so that legacies of political-economic power and inequality, sociohistorical contingencies, institutional trajectories, cultural projections, and the necessities of daily lives help shape how environmental activists and institutions define their social work and encounter the nonenvironmentalists whose behaviors they seek to modify. Furthermore, the targets of environmental initiatives themselves— people who actually live on and work a landscape—introduce and assert specific agendas and knowledge that challenge how this work will take shape, seeking to redefine environmentalism's processes and outcomes to suit their own interests and perspectives. For rural Costa Ricans who often perceived themselves as marginalized by environmentalism's focus on land purchase campaigns and the nongovernmental institutions that were gatekeepers for these processes, this issue raises practical concerns about the credibility and relevance of a social movement that uses a language of sustainability-through-inclusivity. Second, it encourages us to consider how is it possible to make sense of the efforts of individuals or groups outside of and at the margins of environmentalism who seek to intervene and redefine it to fit their own agendas. I deal here with one aspect of this question, asking under what conditions *campesinos* and other rural residents can redefine the scope and meanings of environmentalism to integrate their own interests. Because they are ultimately not reducible to easy solutions—for instance, how is it possible to quantify and manage processes of identity formation and maintenance?—such questions are inherently difficult for activists and institutions to resolve. Yet their significance is that in answering them, we have to address and evaluate the concrete power-infused processes and spaces in which the meanings and relevance of environmentalism and its relationship to regional development take shape, are articulated, and are modified.

The fact is, the various forms of "social work" I explore here demonstrate clearly that there is no singular apparatus of environmental conservation or vision of sustainable development in Monte Verde, but an array of practices and philosophies about how to achieve nature's protection. The three chapters in this section address these concerns by telling stories about three different encounters of "community participation." Each chapter reveals something distinct about the difficulties and opportunities facing environmental activists and institutions trying to define, negotiate, and mediate these processes, as well as the ways people targeted for these interventions assert their own visions of social and natural change. Chapter 4 focuses on what we might call a situation of "nonformal participation," a negotiation between a farmer and environmentalists over a puma that killed some of the farmer's animals. It led to an "outreach" effort of sorts that is not programmed into environmentalist social work efforts because of its irregularity, but that nevertheless confronted activists with some deeply unsettling issues about how they define the participation of others in protecting wildlife and habitats, and how others seek to define their participation on their own terms. Chapter 5 examines the rise and dissolution of an integrated development and conservation project called San Gerardo Project, which coupled a remote rain forest community together with the Monteverde Conservation League in an effort to allow people to keep living in the rain forest instead of ejecting them. This chapter examines how the institutionalization processes of the NGO overcame and redefined an initiative based on "community participation," and how heterogeneous ideas about participation and membership undermined the project. Chapter 6 examines a conflict over the Reserva Santa Elena, a small cloud forest preserve that has been publicized far and wide as a successful story of community involvement and benefit in conservation and ecotourism. It shows the considerable unevenness in how the language of community has been appropriated and employed in this project, and how the label of community, with its connotations of unity and identification with place, has obscured dynamic changes in Monte Verde that are the result of deeply held social rifts and engagement with translocal forces. In sum, these chapters provide an ethnographically rooted evaluation of how certain commonly circulating notions like "participation" and "community involvement" are actually shaped and contested in specific contexts.

# 4 Testing the Boundaries of Environmentalism in a Participatory Age

## Introduction: The Puma in a Tree

Late one afternoon during the mid 1990s, a colleague and I returned from working at the Reserva Santa Elena cloud forest preserve (RSE) as we usually did, walking the six kilometers by dirt road to the village of Santa Elena where we lived. There were no tourists in their rental cars from whom we could hitch a ride—one of the few forms of motorized transportation on what was then a quiet back road (but that is now well-traveled because of the higher number of tourist attractions since the late 1990s)—and we settled into the routine of brisk walk and conversation. Several kilometers ahead we heard the sounds of motorcycles and cars on the road, and our hopes for a ride rose. Within a couple of minutes they sped rapidly by us, heading purposefully in the direction from which we came, toward the RSE. We knew the motorcycle riders, one of whom was a naturalist tour guide and the other a forest guard for the Monteverde Cloud Forest Preserve. Following closely behind them was a Guardia Rural police officer, who sped by in his jeep with a preoccupied look on his face. We wondered where they could be going, especially since the Reserva Santa Elena was closed for the day, and aside from the cloud forest preserve there were only several dairy farms in the area.

Several minutes later, two prominent Santa Elenans pulled up in a Jeep and upon recognizing us, they stopped to greet us. They were both longtime friends of my colleague, and both of them served on the RSE's board of directors. They were excited, even giddy. One of them asked us if we had heard about the *león que estaba encaramado* (the puma that was driven up a tree) on Pedro Solórzano's land, a dairy *finca* several kilometers from where we were, in a remote upland area known as San Gerardo Abajo. We admitted that we had no idea, and they invited us to join them to go see it. In the car, one named Don Eugenio explained that the word around Santa Elena was that a puma had attacked some of Solórzano's farm animals and that his dog chased it up a tree, where it still remained this afternoon. Solórzano had apparently threatened to kill the animal, but had been in contact with the area environmental organizations, and they had been out there all day negotiating with Solórzano to prevent him from killing the cat. For this reason, the excitement in the Jeep had a nervous edge, because they knew the harsh fines and possible prison sentence for Solórzano if he killed the animal. Don Eugenio and his friend were going under the pretense of representing the interests of the RSE, since the cat could have come from there. Don Eugenio admitted to us, however, that he was going to see the animal itself. He explained that their fathers, who were among the original *moradores* (settlers) of the Monte Verde region, viewed pumas and jaguars as common threats. They, on the other hand, all middle-aged men, had not seen any except for those that hunters brought back already dead. Agriculture had expanded so rapidly in the past four decades that these men (themselves farmers) did not have to put up with these difficult visitors. But the word around the region was that the number of encounters was increasing, especially in the upland farms like where we were going, because the populations of large cats were on the rebound due to successes of formal landscape protection.

When we arrived at the entrance to Solórzano's property, a North American man with a long ponytail greeted us, and told us that he would not allow us to enter the *finca,* since the policeman had empowered him to keep people out. I noticed several motorcycles, among them those that had earlier passed us, as well as several trucks that belonged to the two important conservation organizations and landowners in the region, the Monteverde Conservation League (MCL) and the Monteverde Cloud Forest Preserve (MCFP). The man, who wore a shoulder holster with a pistol, was a caretaker for absentee North American landowners who were building a luxury hotel on land next to Solórzano's. He turned to me, and in English declared with solemnity, "Man, this is an age-old story. There's another farmer who's gonna kill a cat, and the world will have one less of these beautiful animals. This is the real thing! I've been calling people all day to get them here so there are at least witnesses if Solórzano kills it. A puma took some of his animals a couple of months ago, and I told him to let it go. I said next time it happens, let's call the environmentalists." I looked over at the naturalist tour guide who passed us on the road. He was unpacking a video camera at his motorcycle, apparently to carry down to the scene where the drama was unfolding. We both looked at him, and the North American man explained without my prompting, "I want some of the guides to be here with their cameras, so if someone kills it, we have a record of it. Pictures are worth a thousand words, you know. Solórzano wants to kill it, and if he doesn't, it'll be that cop who kills it."

Don Eugenio and companions scoffed at the man's presumption that he could prevent our entrance (they had known Solórzano for many years, and besides most landowners here do not mind people dropping in) and we entered the *finca*. As I followed them, the North American called after me, "You might as well go too— it'll be another pair of eyes in case there's trouble." After several minutes walking across pastures and squeezing through barbed-wire fences, we approached a patch of cloud forest, on the edge of which several forest guards were sitting and laughing at each other's jokes. No one was surprised or offended that we had arrived, and after greetings they motioned us to enter the forest where about six other people were milling about, periodically looking up. They too were not surprised we had come, although unlike the joviality on the forest edge, the tension here was palpable. About thirty feet up, sure enough, was the puma, a large male that was nervously peering down on us from a moss-covered tree branch.

There was a feeling of expectation in the assembled crowd, each of us waiting for something to happen. Every now and again, Solórzano would lead people off on one of the foot trails leading out of the forest patch to survey the damage, showing them half-eaten corpses of geese, chickens, and a calf. True to his reputation as a locally famous *monteador* (hunter), the policeman turned to Don Eugenio and in a low voice remarked (wistfully, I thought) that he could hit it using his rifle, as he had done several times before with cats at this proximity. But then he explained in a louder voice that it was against the law for anyone, including himself, to kill a puma. This animal was *patrimonio nacional* (national patrimony), defined as such by wildlife protection laws. This is why he was here, to make sure that the puma would not be killed, and to represent the legal authority of the state if something unfortunate should happen to the cat.

Solórzano had called MCL officials to his farm, he claimed, because "it would hurt me to have to kill the puma, because we aren't here to destroy things." At the same time, however, he felt that someone needed to take responsibility for his losses and to compensate him for not taking revenge on the cat: "we campesinos have been lastimados [damaged] by animals for a long time, and it would set a good example if the environmentalists would help us live with these animals." Later, he would say that he too was a *conservacionista*, sharing their love of wildlife. As I later learned, although he had lived on this farm for only two years, this was only one of several occasions in which he lost farm animals to a large cat like a puma or jaguar. The typical reaction to situations like this in rural Costa Rica, most everybody told me, had always been to kill the cat. But Solórzano called in the MCL because he knew it had paid financial compensations for losses to jaguars and pumas to *campesinos* like him on several occasions in the past. Employees of the MCL were acutely aware of the payments they had made, but they were made when economic times were good. Their association was in an increasingly difficult fiscal crisis as those funds were drying up throughout the mid 1990s. Even though an administrator and forest guards of the MCFP were also there, they claimed they were only there to observe and did not take an active part in the negotiation.[1] "We're just here to be witnesses," an administrator told me, "The League is bringing some help. Hopefully, they'll set it free. It should be taken to a natural area, more wild than here."

After we waited around for another half hour or so, a member of the MCL's board of directors, who is a North American biologist, entered the forest patch with a small entourage of forest guards and an older foreign man that nobody recognized. The board member introduced Solórzano to the man he brought with him, a Swiss man who owns a small zoo in the Guanacaste lowlands several hours away that rehabilitates and maintains large cats and shows them to tourists. A circle formed in the center of the forest patch, around Solórzano and the Swiss man. Recognizing his audience and opportunity, Solórzano histrionically pointed at the cat above, and declared, "I hear you buy pumas. Well, I have one and I want to sell it! You give dollars or *colones*?!" Amidst the laughter that followed, the man responded that he already had four pumas. He added that he was in no position to buy the puma since the Costa Rican government would fine him for illegally trading in wildlife. Solórzano asked him how much he thought the cat was worth. The Swiss man responded that it would be difficult to get it down: even though they had access to a gun that shoots tranquilizer darts, it would not be wise to shoot the cat since it would be hurt by the fall. He added that one option would be to take the cat to another part of the country and let it loose there, but no cage could hold the animal once it awoke, since it would be angry. The Swiss man admonished Solórzano, "You should want the cat to go free since it belongs to no one and it would serve conservation. If you want my counsel, learn to live with it." Again, Solórzano pushed, "It's an expensive cat, but I can give you a good deal!" More laughter.

Solórzano grew quiet as the Swiss man's tone grew slightly more formal and he began to lecture everyone on the behavior of pumas. He questioned whether one animal could have caused all the damage that had been attributed to it. Disinterested by this lecture—who else would have killed these animals?—and visibly annoyed that he had lost control over the audience's attention, Solórzano slipped off. In a quick calculation, the MCL president decided to follow him. It was during this short walk offstage that they worked out a solution to Solórzano's situation, although none of us would know it until the next day. The impromptu lecture eventually petered out as people had fewer questions for the Swiss man, and since it was growing dark, people started to leave. As we walked out, my companions discussed what they had just experienced. Reflecting his view that Solórzano would act as any "normal" *campesino* would act, one man observed,

"That poor animal, he won't save himself. The farmer will just wait until it gets dark and then shoot it." He expected that he would do the same if he found himself in a similar situation. To this, a forest guard who was walking with us reacted, "You know, Solórzano screwed up in this situation. It's obvious that he's trying to make money here, but he's doing it the wrong way. He should have called all the tourism guides and charged them to bring tourists to see the cat. He could have made a small fortune. But he's just a *campesino*, he's simple." In the following days, I would hear other Monte Verdeans repeat this view, that Solórzano missed the opportunity to turn the puma into a tourist attraction.

That evening, an MCL forest guard stayed at the farmhouse to ensure that the cat escaped. Several days later, Solórzano sent formal letters to the MCL, the MCFP, and the RSE (the environmental organizations that were present the previous afternoon), requesting financial compensation for his losses and materials to build secure fencing for his calves and fowl. None of the organizations offered to pay him any compensation, much less (in some cases) even gave him a formal response. As an MCFP administrator explained in an interview several weeks later, this is because the organizations did not necessarily have any policies to deal with situations like these. Preoccupied by the precedents that could be set by compensating this farmer, he reflected,

> Well, we haven't had to confront many situations like this in our institution, something very concrete like this. We have been involved in arrests of hunters. People identify us as working in these things, but in reality, we have never set an institutional policy to deal with situations like what happened on that farm. We know that paying someone an indemnization is a double-edged sword, because people will use it to make business for themselves. They could lose an animal and say it was a jaguar, even if it wasn't. So it's a problem. Even if resources were available for this, it could get very expensive, because sometimes there are a lot of animals.

He added that it is necessary for the different environmental organizations to coordinate a policy to deal with such crises. But a fundamental problem confronting them has been inconsistent inter-institutional coordination. This is based on years of mutual distrust, the product of competing institutional agendas and legal conflicts over landholdings. He concluded that, "The problem here is that there is not a protagonistic element that can coordinate these situations. It should be the government. They manage the law, and they could provide support. But in the meantime, who will take leadership and come up with solutions? No one. That's why this letter was addressed to all of us." In spite of his suggestions that the different organizations develop a joint policy to stand in for a government that generally has not become involved in Monte Verde environmental issues and dilemmas, there never were any formal meetings to discuss possibilities.

Another kind of response, equally aware of precedent, came from the director of the Reserva Santa Elena several days after the letter arrived. During a chance meeting on the road near the Reserva Santa Elena between himself, myself, and several MCL employees who were passing by, he said, "I don't think much of this request for money. Conservation is for everyone to share, so everyone needs to learn how to respect nature. How do we know our investment would be good if we gave him money? Because the puma could return again and he could kill it then, so he makes out with money and the puma is dead. Also, if it continues to kill, he'll continue to ask. Even if we had money to give, we wouldn't give." In other words, he explained in response to a follow-up query, environmental organizations have no particular responsibility toward farmers, since everyone should have the same ethical responsibility to protect nature. Reacting to this statement, one of the MCL employees observed,

> This presents some interesting issues for conservationists to think about. This is a puma. If it were a tick, or a small animal that people don't pay attention to, this wouldn't be an issue. Pumas have more presence than a tick. Solórzano would not have any basis to make claims if he were invaded by ticks, that's part of the bargain being a farmer. For some reason he can make a claim on this species, although we could also argue that this is part of doing business as a farmer.

He followed with a provocative question: why are pumas—"charismatic megafauna" in the language of environmentalism—considered worthy of intervention and protection, while others are not? His question, perhaps a rhetorical one, remained unanswered and everyone went back to their business.

Solórzano did end up receiving assistance for his trouble, but not from any institutional source. Joe Collins, the MCL board member and biologist who resides in the Monte Verde region with his family, offered to help Solórzano pay for the creation of fences to protect his animals. Collins explained that he became involved that afternoon by accident, since as a board member he would not normally have become so personally immersed in the day-to-day operations of the organization. But as he joked with me once, "It's not always clear where the League is as an institution and where I am as a sort of loose-cannon biologist." He explained that part of his reason for becoming involved was that he had a "special bond" (as he called it) with Solórzano, resulting from a situation several years before in which a student from a visiting U.S. university biology course that Collins taught became separated from the rest of the group during a natural history expedition deep in an area rain forest preserve. Solórzano, who happened to be "walking" (hunting, implied Collins) in the area that day, helped the student cross the raging river that separated her from her group. Collins reflected,

So we sort of bonded. I don't hold his lifestyle in contempt, although there are other biologists and conservationists who would see him as a dangerous hunter-type. Anyway, I don't think the whole negotiation over this puma would have happened as it did unless we had this pre-existing positive relationship.... His point was that as long as somebody was willing to help him replace his animals and secure them, he'd be fine. So we came to a general gentleman's agreement, with the details of who would actually pay for the animals and all this stuff being left for later. That, of course, is where I screwed up, since I'm not very good about saying, "Well, the official policy of the League is such and such" and then if it isn't the official policy will follow, or something like that. The thing I think he did understand is that our organization is not the only entity that's interested in having intact fauna around here, so that's why he approached other organizations too. In other words, we're not the last bastion of conservation around here.... On the one hand, we need to put our money where our mouth is. But what happened this day was sort of an invention on the spot. You can imagine with something like a resplendent quetzal leaving a preserve to go onto private property, the individual property owner is helping the quetzal population survive. If a part of the economy here relies on quetzals then you could easily see that many people benefit by having a relatively intact fauna. But it's not hard to justify saying yes we will do what we can to help people with wildlife problems.

Collins had convinced Solórzano to send letters to the other environmental organizations in his appeal for compensation, reflecting his awareness that protecting the forests and its species has brought benefits for certain people at the expense of others. This would be one way of spreading the responsibility for the disparity, in which the cards, so to speak, were increasingly stacked against Solórzano as puma populations rebounded. In the end, though, Collins's association never came to any agreement internally on whether it should pay any compensation to Solórzano, fearful of the precedents it might reinforce with other farmers. Caught in a bind, since no other organizations responded with support—themselves fearing such a costly commitment—Collins drew from his own personal finances to help Solórzano pay for the materials to build fences and cages.

## A Brief Interlude: On Cultural Performance

Long before we convened at Solórzano's *finca,* the political and social boundaries of environmentalism in Monte Verde had taken shape in such a way that Solórzano knew pretty clearly several things. First, as long as he lived where he did, that is, in the fluid border-zones between farming communities and large nature preserves, where mobile species do not respect the arbitrary boundaries of nature preserves, some people would be benefitting at his expense, particularly the North American ecotourism developers that were his neighbors who could "sell" wildlife encounters, as well as the institutions that advocated formal nature protection. Second, he knew whom to approach with his claims, reflecting a widespread perception that environmental groups with international funding connections were wealthy and inclined to interact with people like him in ways that private entrepreneurs like the ecotourism developers would not. Third, that as an outsider to environmentalism—he was neither a member of any environmental organization nor did he have a history of being involved in their formal outreach programs—he knew where his bargaining position lay. Although he did not know exactly what would result from his efforts, he drew from a technique environmental educators often utilize to spread their message—theatrical and performative events—showing his own savvy in creating a theater of persuasion, by showing the visitors half-eaten bodies, declaring himself a *conservationista* too, joking about the price of the cat, and so on. He did it in ways that played up his position as supplicant, while also playing up the implicit threat that he, as an autonomous landowner, can unceremoniously (if surreptitiously) kill the cat once these people leave.

For days afterward, the scene still excited participants, including myself: its emotional intensity—the excitement of seeing an actual puma, the fear of what Solórzano could do to it, the desire to find some way to resolve it in the long term—contributed to the sense that this was an extraordinary incident. Most of us realized that we were witnessing a struggle over something more than just the life of this particular cat, but how Monte Verde's major environmental institutions would address and incorporate this farmer's livelihood concerns. Of course, it must be said that occurrences like this are common across the world, and loom large in discussions about the dilemmas of wildlife protection—wolves attack sheep in the greater Yellowstone ecosystem, coyotes eat family pets in suburbia, and lions gorge on Maasai cattle in the Serengeti. Yet like all those other situations, this story has very particularistic elements—"differences that make a difference" as Levi-Strauss used to say—because its genesis and outcomes are as much products of a particular moment in landscape and environmentalist histories as they are open to diverse meanings and interpretations forged in and beyond the encounter itself.

This situation was neither routine nor a watershed moment for environmentalism in Monte Verde. But it was a specific kind of encounter, a "cultural performance," or a public irruption of tension enmeshed in a play of power where social contradictions that prefigure and stand outside the performance itself surface, are addressed, and perhaps are even reconciled (Turner 1974; Bailey 1996; Parkin 1996). Given a deeper history of ebbing and flowing tensions and alliances between farmers and environmental groups, of which this incident was but one expression, it would be overly simplistic to reduce what Solórzano did to a mere (or crass) search for compensation for his losses. Even if this was an important motivation, this situation was part of a wider, longer, and ongoing dialogue. In fact, he was symbolically contesting widely circulating notions of people like him—poor farmers—as inherently hostile to wildlife, and as actors to be acted upon, not collaborated with, in nature's protection. By declaring himself to be an environmentalist too, he was testing the boundaries of who can credibly identify as one. On one level he succeeded in asserting a certain power, because his meeting obliged people to pay attention to his claims. But interpretations of the performance and its drama were varied, and the effect was not as transformative as might be expected, as external contingencies beyond the performance itself undermined and reshaped the outcomes he sought. And although his performance emphasized the relevance of his issues for environmentalism, he was not able to convince anybody that he was an environmentalist himself, even as he clearly asserted what he considered to be environmentalist principles of respect for wildlife. This is in spite of the fact, as I will show in this chapter and the next two, that what and who is "environmentalist" in Monte Verde have been deeply contested and fluid categories and affiliations. In the end, the way this situation was resolved suggests that much more was going on outside the performance itself that helped give it its final solution, reflecting a particular set of priorities and contingencies in how Monte Verde environmental institutions have defined their social work, the unforeseen involvement of Collins, the various meanings people in positions of influence attributed to Solórzano's motives, and their concern over the precedents indemnization might reinforce.

In important respects this is but one of many face-to-face encounters taking place in people's daily lives in which environmentalism's boundaries have been tested, negotiated, and potentially stretched to address and incorporate pluralistic concerns. The significance of this particular event is that it shows how a particular social form—of orchestrated but open-ended performances wherein certain basic categories and tensions in a status quo are opened up for consideration and critique—has become an important means through which the meanings and practices of environmentalism are shaped in the era of rhetorical inclusivity. As Parkin has observed (1996: xix), the power of the notion of cultural performance is that it "starts from the assumption that human struggle is intrinsic to all forms of social organisation and that peoples seek non-explicit or diversionary, and therefore ceremonialised, ways of resolving the contradiction that comes from having to co-operate with neighbours and kin with whom they necessarily diverge, sometimes fundamentally, in their interests." Attention to such performative encounters encourages us to evaluate how environmental initiatives can (or cannot) acknowledge and accept divergent interests, revealing basic tensions between sustainable development's putative inclusivity and those authorized to define and carry out such initiatives.

# Programming "Social Work"

Before continuing a focused analysis of this situation, it is worth taking another brief detour into a broader discussion of the forms of "social work" that the MCL and MCFP have pursued. It is helpful to consider the contrast between how and why these groups imagine and put into practice certain "outreach" programs, and why a situation like this one—no less crucial in terms of the issues it raises for regional livelihoods, ecological sustainability, inclusivity, and participation—has not been incorporated into any formal programmatic area.[2] Why do some issues and not others become programmatic in environmentalist "social work"? Why have areas like farmer-wildlife relations, which are usually more mundane than this puma incident, involving smaller and more humble but no less agriculturally threatening creatures (coatimundis, raccoons, birds, bats, insects, etc.), not been prioritized through programmatic action?

There are essentially two forms of programmatic social outreach that these two institutions have pursued, both punitive and productive. The former, the punitive, reflects the significant role that both organizations' forest guards play in the buffer zones of the formally protected lands, to prevent poaching, encroachments, squatting, illegal logging, and so on. As the situation on Solórzano's *finca* demonstrates, forest guards tend to be the first on the scene and the last to leave when there is a crisis that involves their institutions. Given that guards are for the most part from the region itself—themselves children of settlers and farmers, and therefore generally at ease in the flow of social interaction in rural communities—it would seem that they would be effective at communicating the imperatives of conservation. And yet, because they are the representatives of what some residents in buffer zone hamlets consider to be organizations that disapprove of their ways of life (such as hunting and collecting in the forests), guards can be considered an unsettling, even threatening presence.

In addition, both the MCL and the MCFP have pursued what I call productive forms of social work, efforts that seek to actively change people's attitudes and behaviors through educational and agricultural initiatives. The goal here is to inculcate values and knowledge that will guide people's actions and produce new kinds of behaviors, so that the punitive threat of policing is ultimately not necessary in the long term. A central aspect of this has been both organizations' environmental education programs, representing a key means of disseminating information on the biological attributes of the protected areas, the importance of formal landscape protection, and the moral and political obligation of residents to actively support the institutions that own the protected areas. But in spite of similar goals, each institution's educational program has had different intellectual and ideological emphases, reflecting distinct institutional histories and priorities.

The Monteverde Cloud Forest Preserve has practiced community engagement through its financial contributions to certain public initiatives, such as road maintenance, a recycling and trash program, ecological and cultural fairs, and grants for schoolchildren (Solórzano and Echevarría 1993: 7). Its most recognizable outreach effort is its small environmental education program, created in 1993. For the past decade, it has employed between one and three people, using funds generated by dedicating 30 percent of the profits from guided tours of the preserve (approximately 6 percent of the organization's annual expenditures, Aylward et al. 1996: 329). Because it has not relied on external donations but on tourism revenues, the program has been able to continue even during a precipitous drop in international funding for Monte Verde environmental initiatives. At the center of its educational mission is a curriculum for elementary-school children (grades four through six) at regional public schools.[3] It teaches a natural history and ecological science curriculum developed in conjunction with educators at Hawk Mountain Sanctuary, Pennsylvania, that is also approved and overseen by the Costa Rican Ministry of Public Education. It also includes, as the one educator in the program explained, "different aspects of the environmental problematic, such as the problem of deforestation, the problem of agriculture, and the positive and negative aspects of tourism." The program's emphasis is unequivocally on natural history and ecology. From the personnel (who themselves tend to hold undergraduate degrees in biology) to the curriculum itself, a central goal of the program is "that the student increases, through different activities, his sensitivity, appreciation, and knowledge of the great diversity of plants and animals that exist in Costa Rica, with special emphasis on the cloud forest" (MCFP 1993).

For the MCL, whose environmental education program preceded the MCFP's—it was established in 1988, but was moribund by the mid 1990s because of internal conflicts and lack of funding (except in La Tigra de San Carlos, on the eastern border of the MCL's vast forest preserve, where one educator labored throughout the mid and late 1990s)—the focus has never been solely on the biological aspects of education. When it was initiated under the energetic and creative leadership of a university-educated Costa Rican man who himself grew up in Monte Verde, the program's goal was to enhance the self-esteem and social prospects of youths in the region by introducing them to practices of sustainability, self-reflection, and cultural reaffirmation. As this man explained, "We were interested in knowing our community, its history, its economic and political organization. It was about raising people's political and cultural consciousness. So we began with the idea that the community was the most important element of conservation, and we explored how our communities were organized to serve it or undermine it. We also had classes and did theater exploring the value of our rural and agricultural traditions, and we created organic agriculture projects in schools." For example, the program sponsored events and activities they called *rescate cultural* ("cultural rescue"); organized campaigns to raise awareness about the growing profusion of trash in the region; sponsored youth theater groups and performances to raise awareness about problems like drug and alcohol abuse; worked through a Catholic youth group to organize a "festival of the river" to celebrate the Guacimal River's significance to the region's ecological, social, and economic life; developed a "periódico comunal" (communal newsletter) and worked with a community radio station ("Radio Monte Verde") to create youth-centered programming; and promoted public analysis and discussion about issues like the declining quality of the region's waters and streams and the dilemmas of tourism development.

Within the MCL itself, such work was celebrated by some but reviled by others. Some biologists and pro-land purchase campaign factions were especially critical, and by 1991, took a stand against the program's leader and what they considered his polarizing politics. Indeed, a North American biologist who was president of the association resigned as a protest against the education program, arguing that it focused on "irrelevant matters" such as cultural rescue and public health issues at the expense of teaching people the importance of the flora and fauna found in the protected areas. With the support of other biologist members of the association, who wrote letters and exercised pressure on administrators, the environmental education program was reorganized with a more central place for teaching about biodiversity in the curriculum, although over time it became clear that neither side was convinced that the new curriculum was really as it should be. The conflict demonstrated a deep ideological divide over not only the content of the program's educational curriculum, but also the form such programming should take. For those advocating an ecology- and natural history-centered curriculum, environmental education was about taking people into the forests with experts to show them what is there. In contrast, for those who had been running the program, the purpose of education was to facilitate collective processes of self-realization and understanding about the conditions of rural life.[4]

What has remained more or less constant throughout these programmatic efforts—certainly in the MCFP, to a slightly lesser extent the MCL—is a commitment to seeing children and youths as the most appropriate focus for educational programming. That is not to say that adults have not sometimes been the objects of educational initiatives, as both organizations have periodically held workshops with parents, teachers, and women in a regional craft cooperative called CASEM. Educators emphasize the long-term nature of the strategy, rationalizing it as shaping an environmentally conscious citizenry for the future, who it is hoped can exercise some short-term positive influences on older people. Of course, as the case of the MCL shows, defining the qualities of that consciousness is a profoundly contested process. Some *campesinos,* especially those that are interested in learning about better productive techniques, concerned about preserving a culture they perceive to be changing rapidly, or simply curious about what is in the forests, consider their absence from such programming to be a regrettable demonstration of a conscious neglect on the part of environmental education programs. But most do not question it: they are either too busy to care, or they draw on a variation of the stereotype that "you can't teach an old dog new tricks." The other side of this stereotype, as I heard from various *campesinos,* is that "young people are more fresh," meaning open to new ideas and practices. Whatever the case, the programs' ideological leanings, as well as their trajectories within their home institutions, should give a clear sense for why Solórzano would not have encountered environmentalism through any formal educational outreach efforts: he was marginal, if not outright irrelevant, to such forms of social work.

But the category Solórzano occupies—a landowning dairy farmer— *has* been the object of significant social work, through the MCL's reforestation program and a later manifestation of it called "Bosques en Fincas" ("Forests on Farms"). In spite of their successes—the reforestation program is widely considered by *campesinos* who benefitted by it to be the most worthy and effective initiative of environmentalism because of its contribution to their productive livelihoods— both of these programs have ebbed and flowed (and ultimately discontinued), depending on international funding, local interest, and broader political economic conditions that affect farming and forestry, such as government incentives.

The first formal reforestation efforts in the Monte Verde region were carried out on a small scale by the Ministry of Agriculture (MAG) and the regional agricultural cooperative, Coope Santa Elena, in the 1980s. When they founded their association in 1986, a key concern of MCL founders was the rehabilitation of the degraded Pacific slope landscape, and so reforestation was begun early in the life of the association, initially with trees grown in the Coope Santa Elena tree nursery. In 1989 the MCL formally established a program and its own tree nursery with funds from the World Wildlife Fund. By the time it was discontinued in 1994, it had succeeded in planting more than half a million trees, involving over 260 farmers in 320 projects (MCL 1994a; Burlingame ibid.: 366). At its height, the reforestation program was a largely autonomous unit within the MCL, and like environmental education, identified more closely with the needs, interests, and perspectives of rural communities and small-scale dairy and coffee farmers than with the direct biodiversity protection concerns of biologists.

In spite of the conservation motivations and benefits—studies showed reforested windbreaks beneficial for bird species (Nielsen and DeRosier 2000) and tree diversity (Harvey 2000)—program workers promoted reforestation as a farmer-driven and production-centered initiative, only indirectly "educational." For farmers, wind has been a determining factor for agriculture and livestock management. Because many farms in the region are small, farmers could not always dedicate large areas of their land to reforestation. But they quickly perceived the benefits of establishing windbreaks, which helped prevent soil erosion and decrease stresses on cows and crops, leading to noticeable increases in productivity. Some dairy farmers, for example, claimed that their cows were producing two to three times what they did before. Windbreaks were also desirable because they provided trees for fence posts, edible fruits, lumber, fuel wood, spring and water protection, and cattle food during dry spells.

Farmers carried considerable responsibility in laying out and planting their windbreaks, appropriate given that they typically had intimate knowledge about local soil conditions, wind patterns, water flow, flora, and fauna. They also had to pay 60 percent of the costs for reforestation. MCL reforestation workers (around fifteen at the program's height) and MAG extension workers provided technical advice throughout the process. Because many farmers could not afford the expense of planting several thousand trees (saplings cost 30 colones each, or roughly 25 cents), they received loans that the government of The Netherlands had provided through the Costa Rican Forest Service (Dirección General Forestal, or General Forestry Divison).[5] Unlike other government incentives and entitlements, where landowners had to show title to their lands, participants (many of whom did not have their ownership officially registered) needed to show only an *escritura,* or bill of sale. Since many farmers had not registered their lands to get title, this allowed for widespread involvement in the program.

During the first years, farmers more often than not chose exotic species, cypress (*Cupressus lusitanica*) and casuarina (*Casuarina equisetifolia*) being most common, because they were quick-growing and on the government's approved list of trees for incentives. But as MCL workers grew more sophisticated in the collection and germination of seeds (and convinced the government to expand its list of acceptable species for incentives), they were able to recommend native species. During its most productive period in the early 1990s, the nursery grew as many as forty-nine species (six exotic and forty-three native species). Two species—Tubú (*Montanoa guatemalensis*) and Colpachí (*Croton niveus*)—were the most popularly requested trees for windbreaks. As species found naturally in the region, they exhibited strong wind and disease resistance, grew fast, and were useful as living fence posts.

Like guards and environmental education workers, MCL reforestation workers were mainly young Costa Rican men from Monte Verde farming families, and so they often bore the brunt of hostility toward their association's land purchases and perceived domination by foreigners. In fact some workers were themselves critical of the land purchase campaigns, contributing to tensions within the MCL. They approached their work not simply as a practical collaboration with farmers, but as an ideologically tinged effort to legitimize the role of *campesinos* in the MCL, and more broadly, Monte Verde's environmental movement. One reforestation employee described the program as one that was based on the "popular" wisdom and knowledge of farmers, adding that "The people are an inexhaustible fountain of knowledge, wisdom, and initiative. *Campesinos* have a lot of experience, they know about what they are doing, and a lot of the forestry engineers and biologists don't recognize that. They often come in with the assumption that the farmer is ignorant, so they think they can do what they want. For us, the fundamental thing was not to impose on the farmer, and everything we did was in accordance with his plans."[6]

In the MCL, there was great enthusiasm and pride in the accomplishments of the program, yet there was deep ambivalence about the formal involvement of farmers in the association. A handful of farmers in the program sought and gained membership. But in 1992, participants in the reforestation program marched on the MCL offices demanding a more substantial voice in association affairs and full membership in the general assembly (Burlingame ibid.). If they were granted membership, farmers would have a majority vote, and could have a major impact on association policies. As I will discuss in the next chapter, there has been a consistent fear among some prominent founders and members that opening membership up to farmers could undermine the association's biodiversity preservation activities (sometimes described as a situation of "the fox in the chicken coop"), and the general assembly rejected the farmers' demands. Instead they set up a "Comité Forestal" (Forestry Committee) to represent the point of view of reforesters to the general assembly and administration, which it did for over a year. This enraged and further alienated reforesters from the association (even as their loyalty to the reforestation program and its workers remained strong), because it demonstrated to them that whatever concerns they might bring to the broader association would be unwelcome.[7] By 1995, the reforestation program ended as the international grants funding windbreaks ran out, and the DGF dissolved its incentive program because of structural adjustment-motivated cuts in government spending. Reforestation efforts continued for several more years under a new program called "Bosques en Fincas," whose main goal was to protect and expand forest fragments outside the preserves, especially in areas where quetzals and other frugivorous birds migrate, though a combination of small-scale reforestation and building fences around forest fragments.

## Toward an Ad-hoc Solution

Although he lived in the buffer zone of Monte Verde's complex of forest preserves, literally within a couple of miles of the MCL's International Children's Rainforest, he would not have necessarily known that the MCL was experiencing declining international support for all of its programs. It is clear that the boundaries of the MCL as an association have had a certain fluidity—farmers could join as individuals, and some did—but the boundaries became rigid and exclusionary when farmers sought to join *en masse*. But the fact is, when this situation occurred, Solórzano was not approaching the same organization that participants in the reforestation programs had confronted in the early 1990s and found ambivalently exclusionary. Only several years later, the MCL had begun to experience a profound association-wide retraction, and had begun worrying less about how to involve people in the buffer zones in its work, and worrying more about how it was going to guard and pay taxes on its large property.

Another aspect of these problems is that within the organization there were internal debates over what some considered its perceived paternalism. Some administrators called this the "Papa Liga" (Father League) syndrome, drawing on a stereotype that Latin Americans are dependent on their states and civil society organizations to organize important public initiatives. In this version, because of its prodigious access to international funds (and the generally minor role of the state in Monte Verde), the MCL became a kind of surrogate state, throwing money around in such a way that it created an economic and political dependency on itself for conservation initiatives. Ironically—and this is what concerned these critics—this contradicts and undermines the MCL's very efforts to achieve sustainability by engendering autonomous action in support of conservation. This perspective reflects a sense (in some senses justifiable, but also debatable) that many MCL employees and members have held, that it is *the* central organizer of environmental initiatives in the region, and that what the other organizations do is subsidiary or peripheral to the MCL's central status. This paternalism seemed especially acute after the end of the reforestation program, because with its end most farmers simply quit reforesting even though some declared their desire to continue windbreak reforestation. The concern was that this indicated the MCL had not succeeded in cultivating an autonomous reforestation ethic among farmers, and in fact it had created such a strong reliance on the materials and technical advice it provided that farmers would not or could not do it on their own. The concern over paternalism also extended to the situation on Solórzano's farm, specifically that the association's past willingness to compensate farmers who lost animals to large cats had created an undesirable dependence on the MCL. Why is it not just the "price of doing business" as a farmer? Because, it was thought, the MCL had created a dependent population who turned to it for things that it would not do or suffer on its own. At its extreme, this is a cynical position that views the targets of the association's social work as passive except when they can exploit its largesse.

Such thinking overplays both the association's largesse and the passivity of those people who live in buffer zones. Indeed, Solórzano demonstrated an interesting ability to turn a struggle over a puma's life into a bigger question about what environmentalists can do to help a farmer, and who can credibly claim to be an environmentalist. This was not something that Solórzano meticulously planned— there were simply too many contingencies, from Collins's involvement to the arrival of MCFP officials, not to mention the arrival of my contingent of RSE officials—but he did know how to take advantage of the situation. In fact, it was he who turned the forest patch into a theater of cultural performance, because he was able to demonstrate in a compelling way for more or less everyone present that his interests were in direct competition with those of his neighbors. It is here where the meanings of his claims to be *conservacionista* take on their significance. His was not a statement of similarity more than it was a statement of differentiation from the various institutions and their representatives. He was not suggesting that his approach to the natural world would be the same as theirs. Indeed, his ideas about nature differ from at least some of those present. Take, for example, the MCFP administrator, who told me that he would prefer that the puma be taken "to a natural area, somewhere more wild than here." At a later point, I mentioned this to Solórzano, whose response was a simple "but they [pumas] don't care about that." Put in more formal language, Solórzano was challenging any static and bounded notion of nature, asserting that the relative quantity of nature, as in "more natural," has little relevance in his world.

The fact is, Solórzano was also not confronting any singular apparatus of environmental conservation or vision of sustainable development that day, but an array of practices and philosophies about how to achieve nature's protection. The representatives of institutions that were present—the MCL, MCFP, RSE, the Guardia Rural—have each had distinct visions and niches within the broader commitment to landscape and wildlife protection in Monte Verde. They also each have had certain relationships with donors and home bureaucracies that affect how they approach an issue like this. One constricting factor is that donors, when they have been interested in issues of "sustainable development," have tended to support formalized programs like environmental education and reforestation, and not explicitly target such issues as farmer-wildlife conflicts.

Each institution also had specific (and internally contested) philosophies of who should be the target of their social work initiatives, and how to work with them. For the police officer, whose role in regional environmental issues and conflicts has been generally quite limited except when he processes offenders that forest guards catch, the goal was to enforce wildlife protection laws, even though everyone present knew that his personal commitment to these laws was suspect. Because its profits went to financially support the cash-strapped high school in Santa Elena as a means to improve the quality of secondary education in the region, there was little chance the RSE would contribute, and in fact Solórzano did not originally invite them because he did not identify them as a politically viable source of financial support. The MCFP, which has declared its successes in achieving sustainable development (Aylward et al. ibid.), had no real intention of negotiating with Solórzano, largely because with a few minor exceptions the scope of its social work has not gone far beyond teaching natural history to local schoolchildren and protecting its ownership over its preserve with forest guards. Its definition of sustainable development focuses on generating enough financial support to protect its forest preserve, largely through ecotourism, and there is little doubt that it has been doing a profitable business as a tourist destination.

In fact, after leaving the scene that afternoon and talking about it with a number of the people who were present, two relatively different views of Solórzano's situation emerged. The first was that Solórzano was engaged in a disingenuous ploy to get money, since given the opportunity he would probably just kill the puma anyway. This became apparent upon leaving the *finca*, when one of my companions pointed out that the puma was doomed whatever the financial outcome, since Solórzano would simply act as any other small farmer would. From this perspective, Solórzano's claims to be environmentalist were just a cynical manipulation whose goal was to get money. The other was that Solórzano was doomed to fail in his negotiation because he did not properly understand the nature of the business opportunity presented to him. Several people told me that he completely missed the fact that he had a tourism attraction on his lands, and that naturalist tourism guides would have gladly taken paying customers there to view the animal. The conclusion was that he was but a simple *campesino* who would not understand the sophisticated workings of the tourism market.

To say the least, the diversity of these commentaries demonstrates the impossibility of closed readings, much less common understandings, of the performance that took place at Solórzano's place. But I also think they fundamentally misunderstand the significance of the claims that Solórzano made on environmentalism's institutional gatekeepers. His goal was not to "join" any of these organizations in any formal sense, much less commit himself to their missions, but he was in a position to remind the representatives of environmental organizations that their actions and practices have impacts on the livelihoods of farmers who have a different relationship with the landscape, and this at a time when these organizations either never had or were retracting from their relations in the buffer zones with people like Solórzano. Here was a possibility to (re)open negotiation over what kind of relationship is possible between them given such divergent interests, revealing the possibility for a common space where different approaches to land stewardship could communicate with each other. The fact that it was ultimately settled in financial terms was a pragmatic solution that suited Solórzano (he certainly did see this as a financial issue), and I know that Solórzano was not proposing a long-term series of formal meetings to discuss these things. What is even more important is that the situation revealed an unwillingness of environmentalism's institutional gatekeepers to deal seriously with his claims about divergent interests. In testing the boundaries of environmentalism, he may have discovered that there are multiple versions of it circulating within Monte Verde, even within institutions themselves, but he also discovered that the gatekeepers have considerable power to delegitimate his concerns by choosing not to address them.

But one individual who was willing to take Solórzano more or less on his own terms was the MCL's biologist-board member, Collins. To an important extent, it reflects Collins's own evolving environmental philosophy, which has grown to understand the importance of dialogue across such differences, the importance of paying for the "privilege" of conservation in a country with pressing issues of land distribution, and an acknowledgement that nature is not truly boundable. This is not to suggest that Collins felt that he or his organization "owned" the puma since it may have benefitted by their actions to fight poaching and save its habitat, but that there is a pragmatic obligation to address the long-term viability of wilderness protection models based on static apprehensions of landscapes. But we can also overplay Collins's own philosophical agreement with Solórzano about how to deal with the problems of wildlife in the buffer zones.

Solórzano and Collins did not necessarily have the same concepts about how nature and people relate to each other in the fluid borderlands between wilderness areas and farming communities. To Collins the biologist, the abstraction of natural dynamics in a buffer zone is not necessarily the same as Solórzano's concrete experience of living there. Furthermore, it is necessary to consider Collins's involvement in terms of the personal obligation he felt to Solórzano, born of unusual circumstances, that he might not have felt to any other farmers. Not long after, Collins invited Solórzano and his family on a trip to see the small zoo in Guanacaste where the Swiss man came from, partly out of an expression of goodwill toward Solórzano, and partly because he thought it would be important for them to see these animals up close, to better appreciate them. For an MCL board member to become involved in such a crisis and ongoing relationship was not necessarily unusual—many board members over the years have had the kind of job flexibility that allows them to dedicate time to contingencies facing their association—but Collins emphasized his involvement as a "loose cannon biologist" more than as a representative of the MCL. The fact that the lines regularly blur between individual identity and institutional representation (and not for Collins alone) risked committing the MCL to something administrators did not want, and in the end, it was as a biologist and neighbor—not as an official act of the association—that Collins ponied up the money to help Solórzano build fences. But this was also a pragmatic response to the fact that Solórzano's entreaties to the gatekeepers of funding—the environmental institutions he sent letters to—failed to respond. Collins believed that he had to make an individual commitment to follow through with the support, or he risked both a compromise in his personal relationship with the man and a broader loss of a potential ally for wildlife protection in the future.

What is significant here is that neither Collins nor Solórzano had to actually cross the boundaries of environmentalism for both of them to recognize that as long as people like Solórzano live in that space, some (like Solórzano's American neighbors setting up an eco-hotel) will be benefitting at the expense of others.

But that each interprets this situation distinctly is just as significant. In Collins's case, he recognized, perhaps uncharacteristically given the inaction of others, that he and other environmental activists have important obligations toward those who live in the buffer zone, and should exercise these obligations through their authoritative position and access to resources. Solórzano, on the other hand, sees it as the opposite: he is here to stay and environmentalists must accommodate themselves to the demands of a situation that they helped create. In either case, the boundaries of environmentalism did not move, leaving us with a key unanswered question about the putative inclusivity in the age of participatory sustainable development: are there ways to cross those boundaries? In the next chapter, I consider in substantial detail an alternative model of participation that brought the MCL together with a rain forest community to see if they could develop an integrated conservation and development project.

## Notes

1. MCFP and MCL forest guards share a radio frequency, so MCFP guards found out about the situation over the radio. Solórzano did not call the MCFP to his farm.
2. Note that I do not include the Reserva Santa Elena in this list, even though it too received a letter. The RSE was never a serious option for Solórzano and he probably knew it, since being owned and operated by the generally impoverished local high school, it has a very different status than the other two, which have long been identified with strong ties to foreign funding agencies. For more discussion on the RSE, see Chapter Six.
3. During the late 1990s, the number of schools was fifteen. By 2004, it was reduced to seven.
4. Although it closed as a formal program in 1995 when funding ended, the environmental education program lived on in a kind of twilight state, supporting a youth group called ANJEM (Asociación de Niños y Jovenes Ecológicos de Monte Verde, or Monte Verde Association of Ecological Children and Youth). This group carried on the original vision of the program's founder: to be a venue where rural youths could encounter one another, share experiences, and learn about the social, ecological, political, and economic problems they faced.
5. If, after three years, the farmer could show evidence that the trees were healthy and well-cared for, the debt would be forgiven.
6. This collaboration and mutual respect was celebrated in an annual "Dia del Reforestador" (Reforesters' Day) where as many as six-hundred people turned out to visit model reforestation projects, share experiences, and socialize with friends.
7. In 1996, the MCL created an internal committee on participation, with the goal of seeking broader community support and involvement in its mission. One of the underlying realities of the committee's work was this history of alienating a potentially strong constituency.

# 5  Dismembering San Gerardo: A Cautionary Tale of "Sustainable Development"

Before its dissolution in the mid 1990s, the San Gerardo Socio-Biotic Community Project was celebrated as a model integrated conservation and development project based on community participation. Designed and enthusiastically advertised as a joint undertaking between the Monteverde Conservation League and a settlement of cattle ranchers and small-scale farmers living in the Atlantic slope area of San Gerardo Arriba, the project's plan called for community members to develop organic agriculture and livestock production, produce clean power, support selective high-end ecotourism, and construct a biological station (Boll 2000). This would take place in an area of stunning natural beauty and biodiversity that, in spite of over forty years of *campesino* settlement, still sustained large areas of primary rain forest. According to project descriptions, "The purpose of the San Gerardo Project is to allow these people to live within the protected forest and make a living that is compatible with the rules of the Arenal-Monteverde Protected zone and the philosophy of the League concerning preservation and enhancement of the rain forest." As an MCL worker once earnestly told me: "This has to work, for it has global implications. We are dissolving the boundaries between parks and people."

For the MCL, involvement in the San Gerardo Project initially represented a self-conscious deemphasis of their organizational strategy of coordinating land purchases for formal protection, and a deliberate move toward collaboration based on themes of community and economic "sustainable development." Within several years of these optimistic beginnings, however, most of the sixteen families who had begun to plan and implement the "socio-biotic community" with the MCL had sold their lands to the association and resigned from the project. By the mid 1990s, MCL employees working on the project had only three families with which to work, although there was a newly constructed biological station that brought with it hopes for the project's turnaround. Finally, concerned that the original project had become moribund beyond repair, administrators decided to formally cancel it in 1995, and pursue compensation payments to the three remaining families for their material and labor investments in the station.

While some Monte Verde environmental activists view this outcome as a triumph of their commitment to habitat protection without the threat of human settlement in the fragile ecosystems, others have viewed it as (one told me) "an important learning process" for those environmental activists accustomed to paradigms of diffidence, even hostility, toward rural communities. For still others, however, the San Gerardo Project was a painful failure that demonstrated the incommensurability of planned rural economic development and habitat preservation, and the sometimes creeping, other times dramatically swift, governmentalizing tendencies of a "*non-* governmental" institution. Perhaps most importantly, the replacement of *campesino* communities by tourists and biologists, and the centralization of landholdings and project investments in an NGO, reflect and further the processes by which Monte Verde landscapes have been resignified as spaces of scientific research and touristic consumption, while reinforcing the notion that rural Costa Ricans continue to represent an undesirable impact on those landscapes.

By now, the San Gerardo Project is a bitter memory of what-could-have-been, a biological station that sits at the end of an incredibly muddy road. The area is more or less off-limits to visitors, except for those who stay at the station for research, volunteerism, or corporate retreats. Debates over it have receded far into the background and, in fact, very few visitors to Monte Verde will ever hear about it. But the legacy of its planning and initial efforts at implementation have persisted during the past decade in the skepticism some people have toward the good intentions of environmental activism, or as a check on hubris within the MCL.

More importantly, for the purposes of this section of the book, the San Gerardo story raises important and undertreated concerns about how the community was "dismembered," shedding light on the role that one NGO came to play in redefining this sustainable development project. This chapter seeks to explain why this project did not achieve what many of its planners and members intended for it, focusing in particular on the negotiations and conflicts that took place between San Gerardo landowners, project members, and MCL administrators. Like other anthropological analyses that have examined what James Ferguson calls "the complex relation between the intentionality of planning and the strategic intelligibility of outcomes" (Ferguson 1994: 20; Appfel-Marglin and Marglin 1990; Escobar 1995; Fisher, W. 1995; Leach and Mearns 1996; Fisher, W. 1997; Weisgrau 1997), this chapter raises critical perspectives on both the popular and scholarly tendencies to perceive environmental NGOs as neutral, nontechnical, even effective, intermediaries of social and economic processes related to "sustainability."

There is no single factor or event that resulted in the dismemberment—both the literal unmembering *and* the strategic forgetting—of the San Gerardo Project. Rather, certain sociopolitical and administrative processes, some unexpected, others overlapping, combined to result in the formal dissolution of the project and the fact that its story is more or less forgotten in mainstream stories of conservation success in Monte Verde. Importantly, competing definitions of who should participate in the project and how they should do so, from the planning stages to implementation, contributed to persistent ambiguities in the project. But this chapter is not a postmortem dissection whose goal is to identify the inabilities of particular individuals to "make a project work," focusing instead on the social practices of project design and implementation within a scene of contested perspectives and knowledges on participatory sustainable community development. As such, it is meant as a re-membering and as a cautionary tale to correct various hubristic assumptions that have been circulating during the past decade about the ability of non-governmental organizations to be effective in facilitating participatory environmental processes (Fisher, J. 1993; Fisher, W. 1997).

There is a fundamentally polysemous character of the notion of participation, and when NGOs involve themselves as technical intermediaries of what terms like participation will mean—even as, I will show, they are internally conflicted over how to define them themselves—there can be significant complications in the elaboration of projects, and more importantly, in the collaborations with the people who become targeted for such projects. The reasons why and how NGOs often come to play such a powerful role relate not only to their current privileged status as channelers of international aid and national policies in a context of neoliberal reforms, but to the historical processes of institutionalization specific to the organizations themselves. As NGOs and other grassroots organizations attempt to institutionalize participatory sustainable development, they often engage in a contradictory project in which the priorities and interests of the institution as a whole can eclipse the heterogeneous concerns of its members and target peoples, resulting in clear challenges to the very integrity and autonomy of the communities they intend to help.

# Reflections on NGOs and Sustainable Development in Costa Rica

By now, the comparative advantages of NGO involvement in international development and environmental politics are well-established, and they have been fully integrated into the rhetoric of the development industry and all of its subsidiary enterprises, including "sustainable development." NGOs, whether they are small-scale voluntary grassroots organizations run by volunteers or transnational non-profit corporations controlled by professionals, have been widely heralded for their ability to bypass ineffective or venal government and aid bureaucracies, deliver relief services, promote economic growth, and support participatory democracy and civil society (Korten 1990; Carroll, T. 1992; Fisher, J. ibid.; Wapner ibid.). It generally follows that small-scale NGOs are flexible, experimental, supportive of decentralized initiatives, concerned with processes rather than outcomes, capable of helping people adapt to modernization, and more participatory (Annis 1988; Chambers 1993; MacDonald 1995: 226, n.3; Fisher, W. ibid.; UNNGLS 1997).

These reductionistic stereotypes, especially that small-scale grassroots NGOs represent neutral alternatives to the weaknesses of official development bureaucracies and corrupt Southern governments, have recently come under critical scrutiny as the apparent conditions of people's lives either have not changed or have deteriorated, and as journalists and scholars have begun to document the histories, operations, and effects of specific institutions (Fisher, W. 1995; Macdonald ibid.; Edwards and Hulme 1996; Fisher, W. 1997; Macdonald 1997; Weisgrau 1997). The result is that the construction of NGOs as inherently separate from markets, independent of donor agendas, and as apolitical obscures the normative roles that many have played as agents in extending capitalist markets and implementing certain universal definitions of progress, needs, democracy, and so on. Indeed, as Rahnema (1990) has pointed out, the privileging of grassroots and participatory initiatives, so pleasing in their attention to common people's self-determination, has served as a cover for the increasing privatization of control over public goods and processes.

As an interventionist and benefactor state, the Costa Rican government historically took a central (and centralizing) role in organizing economic development and social measures such as social security and health care (Rovira Mas 1989). Nonetheless, international and grassroots NGOs have played an increasingly important role in the Costa Rican politics of development and its recasting in terms of "sustainability," as mediators and channelers of competing multilateral donors and regional governments promoting these ideals and programs (MacDonald 1995). The Costa Rican field has also demonstrated the general patterns that Korten (1990) identifies as the evolution of NGO orientations: relief and welfare, self-reliant local development, and most recently sustainable systems development. Currently in Costa Rica, sustainable development NGOs number in the hundreds, each operating in highly specific niches, campaigns, and projects, and with distinctive relations to states, donors, scientific knowledge, and specific social communities. The dynamism of the NGO sector in Costa Rica is related to the government's contradictory position with respect to the environment, a theme explored in chapter 2. Capitalizing on the inconsistencies generated by a neoliberal state dedicated to ecologically destructive forms of agricultural development to service exploding foreign debts, even while vigorously pursuing a national park system, NGOs have become directly involved in the provision of services to rural communities, precisely because the Costa Rican state has not been in a position to do so (Rodriguez, S. 1991: 88–9). In fact, the Costa Rican state, in declaring itself open for business as a "laboratory" of "desarollo en alianza con la naturaleza" (development in alliance with nature) during the decade of the 1990s (UNED/INBio 1994), has facilitated the shift in control over sustainability-oriented development initiatives to the NGO sector.

The increasing involvement of environmental and sustainability-oriented NGOs in the provision of services through community and economic development also corresponds to a shift in transnational environmental political discourses on nature protection that emphasizes the management and alleviation of poverty as integral to efforts at nature protection. Since the early 1970s, some Costa Rican environmental activists have argued along these lines: that the protection of Costa Rica's natural resources depends on the integration of rational social management and "ecodevelopmental" practices within the Costa Rican state's efforts at community development and nature protection (Budowski 1985; Fournier 1991; Romero 1991; Mora 1993).[1] In the 1970s and the 1980s, a number of transnational environmental organizations like the IUCN and World Wildlife Fund also began to recognize that the strategy of saving particular species by setting up nature preserves did not treat the root problem of habitat destruction, and made significant investments in Costa Rica (IUCN 1980; WWF 1988; Tangley 1988; Adams 1990).

As transnational corporate environmental NGOs and donors adopted a localistic approach to environmental politics and redefined their mission as promoting community economic development, they and their surrogates have also entered into realms of governmentality traditionally claimed or controlled by governments. While this fact does not mean their intention is to replace states, their involvement has at times undeniably reinforced a shift in relations of power and the provision of services from states to specific localities, and even more so, to their institutional intermediaries (Wapner ibid.). An important strategy has been to rely upon partnerships with smaller-scale, often self-identified "grassroots" organizations to help them define, implement, and sustain their projects, and in fact, the very existence of an organization like the MCL is the product of this dynamic. A key aspect of this has also been to cultivate the participation of local people in the identification of their own needs and the provision of services.

When it appeared in development discourse in the 1950s, participation was meant to end top-down developer-driven strategies and has been positively referred to as the free and uncoerced engagement of "target peoples" in the identification and treatment of their economic and social needs (Rahnema 1992). Notwithstanding Rahnema's valuable observations that "participation" is code for neoliberal privatization schemes and is often imposed on people who have no interest in participating, one of the most problematic aspects of the history of the participation concept has been its universalistic and ahistorical premise that participation exists uniformly in all places and times despite factors of cultural, ethnic, gender, or class difference. In fact, as has been well-documented, non-Western and indigenous peoples, women, and the poor (to name but a few who have been consistently targeted for participatory programs because of their "underdevelopment") have distinctive criteria and meanings for their involvement in development programs and processes of social transformation, based on differing and competing notions of their priorities and needs (Martinez-Alier 1990; Morgan ibid.; Weisgrau ibid.; Kabeer 1994; Guha 1997b; Guha and Martinez-Alier 1997; Esteva and Prakash 1998; Guha 2000). In connecting these insights to the intermediary role of environmental NGOs in Monte Verde sustainable development, the point is not only that diverse meanings of participation exist within the populations targeted for help, but that also they exist within the NGOs themselves. As I will demonstrate below, the lack of consensus within the NGO is both a reflection of and intensified by the cultural, class, gender, political, and philosophical diversity of those who have claimed membership in the association.

Furthermore, processes of institutional growth have significant bearing on the actual meanings and effects of participation. As a grassroots organization scales up and enters a new area of action such as "sustainable community development" with the support of international donors, not only does it pursue new and complex relationships with target peoples, but it must also pay attention to the concerns of donors, as well as to its own survival and evolution as an institution. If the main concerns of the employees include the expansion of the institution and maintaining positive relations with donors, then the best intentions and goals of specific projects can be overshadowed, not to mention undermined by donor interests or processes of institutionalization themselves. These processes include legal and administrative consolidation, infrastructure development and improvement, employee capacitation, and so on. So despite an institution-wide rhetorical commitment to communal participation in the design and implementation of projects, NGO-mediated operations can assume a contradictory process in which daily administrative functions, decision-making, and fundraising fall increasingly within the purview of the institution, since it generally has more access to donors and resources to carry on the project than the target population. The next sections of this chapter explore these themes in detail.

## Planning for a "Socio-Biotic Community" in the Arenal Watershed

For all intents and purposes, the community once known as San Gerardo Arriba exists only in name, and areas once cleared for homesteads have been steadily re-engulfed by forest. From its beginnings in the 1940s, the remote and forested San Gerardo area (approximately ten kilometers from Santa Elena, on the Atlantic slope of the Continental Divide) never sustained a large population, and in fact probably never exceeded twenty-five *campesino* families. Although the land was considered highly fertile and a small church was built, *campesinos* did not move into the area in large numbers because of high rainfall, difficulty of access, absence of a local market for goods, and lack of basic social services such as a school or health clinic. The few who lived full time in San Gerardo Arriba practiced subsistence and small-scale market agriculture and livestock production.

During the early 1970s, the Costa Rican Electricity Institute (ICE) announced plans for a hydrodevelopment project in nearby Arenal, identifying the forested areas south of Arenal (including San Gerardo Arriba) as key watersheds for the dam and lake, as discussed in chapter 2. Within the Arenal National Electric Energy Reserve, restrictive measures were imposed on landowners, including the prohibition of bank credits, denial of land titles, prohibition on cutting trees and burning, and the abandonment of government support for improving roads. Government officials assured landowners that their lands would be expropriated, but it soon became clear that budgetary constraints imposed by the 1980s debt crisis would prevent this from happening.

As state forest guards became more vigilant in enforcing reserve restrictions on cutting trees, residents of San Gerardo Arriba found it increasingly difficult to continue expansive agricultural production. By the mid 1980s, frustrated that they had still not reached land claim solutions with the government or failed to improve even the most basic social services after almost ten years, residents organized the *Comité Pro-Defensa del Campesino* (Pro-Campesino Defense Committee), a committee that with the support of the national small farmer union UPANacional (Unión de Pequeños Agricultores Nacional) pressured the government through national and regional political channels to settle their claims.

The pressures of residents and UPANacional, as well as the creation of the Ministry of Natural Resources (MIRENEM) and its organization into regional conservation units (*unidades regionales de conservación*) under the government of Oscar Arias (1986–1990), led to a renewed interest on the part of the government to secure the watershed's absolute protection. During this period of renewed negotiations, sixteen of San Gerardo Arriba's landowners decided to seek official permission to continue living in the area, proposing to live in the watershed indefinitely utilizing intensive (as opposed to expansive) forms of livestock and agricultural production, with the idea of someday possibly inviting tourists to the area. Their proposal was heard sympathetically at MAG (Ministry of Agriculture and Livestock), but MIRENEM, the agency actually overseeing the situation in the watershed, did not respond, and so landowners engaged in selective tree cutting as a way to gain attention to their claims.

Officials at MIRENEM invited The Monteverde Conservation League (MCL), whose members had been raising funds and purchasing lands for the nearby Peñas Blancas Valley since 1986, to help resolve the dispute between the government and residents of the Reserva Forestal Arenal. Knowing the government's constraints in a time of structural adjustments, they believed that the MCL could draw on its international fundraising connections to raise money for indemnizations (DGF 1987). At the same time, there was reluctance on the part of members of the *Comité Pro-Defensa del Campesino* to accept the involvement of the MCL, because of its identification with land purchase campaigns. Members of the MCL were also divided over whether to become embroiled in the conflictive situation in San Gerardo, expressing concerns over how this might reorient the organization's work and mission.

But faced with the pressure of landowners threatening to cut trees (something that spurred them into action with the Peñas Blancas Valley), MCL administrators decided to formally enter the San Gerardo issue. They invited two University of Costa Rica architecture and planning students who proposed to design a community predicated on principles of ecosystem preservation and economic self-sufficiency, and were searching for a community with which to work. It is here that the idea of "socio-biotic community" was ostensibly born. According to the students' initial vision, the "socio-biotic" would be pursued through three axes of reflection and action: environmental protection (*protección ambiental*), organic forms of agriculture (*agricultura orgánica*), and processes of social enrichment through communal participation (*participación comunal*). At this stage, the scheme was based on a more or less top-down managerial approach to community economic, social, and natural processes, although they pushed for "communal participation" in the planning and implementation phases, asserting that the immediate project priorities had to be identified by the people of San Gerardo themselves, since the quotidian aspects of the project would rest over the long term on their shoulders as much as, if not more than, the MCL's. The students and the several MCL employees (from the environmental education program) who were working on the project elicited and refined these priorities through small workshops, participant-observation research, and field trips to other communities dealing with similar issues.

In the meetings and encounters that took place during 1990–91, the high points in project planning, there was general agreement among San Gerardo Arriba residents, among them men and women who worked as ranchers and subsistence farmers, schoolteachers, and members of the *Comité Pro-Defensa del Campesino*, that their most immediate needs revolved around establishing an economic base that would allow them to continue living in San Gerardo Arriba. Key foci included improving access roads to get products in and out, establishing an organic garden and dairy farm, and exploring the options to bring ecotourists into San Gerardo who, it was considered, would be interested in a living example of sustainable development. Some residents eagerly pursued English and natural history lessons so they could become guides. MCL employees and the architecture students also encouraged the involvement of scientists, who could undertake research on the area's natural history and agroforestry, in order to create a biological inventory and more effectively plan for land use.

The students wrote up the pilot plan, a 179-page document that began to be referred to as the "master plan," and submitted it as a thesis for their architecture degrees (Montero and Zarate 1991). Its final recommendation was to divide the project into three main foci: organic agriculture projects, ecotourism development, and support and infrastructure for scientific research. It was agreed that agricultural production and dairy farming would be in the pastures in the upper parts of San Gerardo Arriba, on the working landscape nearest the road to Santa Elena, which would make it easy to get products in and out. Lands in the lower reaches of San Gerardo Arriba, which are the most remote areas from human settlement (and the most forested), would be areas for ecotourism and scientific research. Optimism was high that the project would work, and two MCL forest guards, only one of whom actually owned land in San Gerardo, decided to join the San Gerardo group. Although they had not legally formalized themselves as an association, participants had begun referring to themselves as "*socios*," or members.

## Competing Perspectives on Participation and Development

But as this optimism strengthened, the MCL pursued its intention to purchase lands for absolute protection in San Gerardo Arriba. In August 1990, the vice-minister of MIRENEM, who had pushed for the absolute protection of the watershed, offered the MCL formal authorization to buy the lands of those families in San Gerardo Arriba who wanted to sell. The MCL began buying lands in early 1991, and most of the more than twenty families that owned land in San Gerardo took advantage of the opportunity. Originally as many as sixteen families had expressed a desire to continue living in San Gerardo Arriba and had become involved in the planning of the project. But with the end of land purchases, only five families decided to remain as *socios* (members) of the project, none of whom had their primary residences in San Gerardo.

As in other regional land purchase campaigns, people sold their lands and left San Gerardo for multiple reasons. The most common explanation of former landowners is that after a long struggle with the government, selling the land was a difficult opportunity to pass since it might never arise again (Vieto and Valverde 1996). For some, it seemed risky to continue with the socio-biotic community with so many neighbors selling out. Still others expressed displeasure with the progress of the project, since few concrete projects had yet been realized when land purchases commenced, or as one landowner complained to me, "We could not continue to attend meetings all the time. We still had to make a living." This willingness to sell created a dilemma for MCL administrators, who were concerned that they would be accused of pressuring people to sell, undermining the project, or contributing to the inflation of land prices in the area. But at the same time, it had access to substantial sums of money that, once allocated, needed to be spent quickly and for the purpose for which it was allocated (as specified by the donors). This dynamic was the result of the MCL's involvement in one of Costa Rica's first debt-for-nature swaps organized by WWF, Nature Conservancy, and Rainforest Alliance.[2]

Although they had spent several years talking over and planning for the San Gerardo Project, it was clear that neither the remaining families nor the MCL had clear agreement on what the role of the other was in the elaboration of the project. Once land purchases began, the MCL's status as the new major landowner in San Gerardo Arriba generated uncertainty among the families, raising concern over how the MCL would actually relate as a neighbor to people pursuing landscape-modifying projects on lands that it wanted for absolute protection. Although the Project's (now) "master plan" identified the MCL as the main supervisor and channeler of funds and technology, what would this mean on a daily basis and how would expenses actually be divided?

It must also be said that the *socios* of the San Gerardo Project had varying commitments to agriculture, forest preservation, the project itself, and even the other remaining families. Significantly, most have never spoken of their involvement in San Gerardo Project in terms that the students and the MCL used, the bureaucratic language of "community participation" (*participación comunal*). Rather, they viewed their involvement in terms of the *inversión* (financial, labor, and emotional investment) and *apoyo* (support) they made and their role as *socios* (members) with certain informal rights and responsibilities. These concepts had distinctive nuances and implications that led to tense relations at times with the MCL, whose members and employees often held different assumptions about their own involvement as paid employees. But there was also no agreement between San Gerardo Project *socios* on the meanings of these terms, which in turn produced tensions between the families involved in the Project.

For example, there was consistent tension between the two ex-MCL forest guards and the three other families, since the guards were seen to be opportunistically capitalizing on their relationship with the MCL to join the San Gerardo Project. According to a man who counts these tensions as one of his reasons for pulling out of the Project in 1993, the ex-guards became *socios* of the Project for the wrong reasons and did not share the same philosophy of *apoyo* for the Project: "They thought they could make money, and so the work they did was on the biological station [see below], which the League paid them for. I always told them that this wasn't proper *apoyo* [support] for our project, that we should work on many things, not just the station, and their work had to be *ad honorem* [unpaid]." Another *socio* explained to me that there was a difference in *mentalidad* (mentality): as former employees of the MCL, they did not show any *iniciativa* (initiative) to take on new responsibilities, responding only to the directives of MCL administrators. The other *socios,* however, had never been employees of the MCL, and vigorously resisted any implications that they ever were.

Adding to these differences in *apoyo* and *iniciativa* were conflicting views among *socios* about the role of the MCL in the Project. On the one hand, *socios* had high expectations, ranging from anticipation of technical support to its leadership role in their educational and community processes. For example, *socios* expected the MCL to channel knowledge and technologies related to organic agriculture, and although it had no specialists in the area, the MCL did arrange for foreign volunteers to work on organic gardens. Moreover, some expected the MCL to validate and manage the processes of group cohesion. According to one of the ex-forest guards who participated as a *socio,* "The principal problem was that the community was not being fortified, and the League wasn't putting effort or money into it. They didn't seem interested in our community dynamic, which couldn't exist until we lived in San Gerardo."

On the other hand, the *socios* were highly sensitive about maintaining their financial independence of the MCL and their political autonomy as a group, which had been forged in years of conflict with the government. According to one woman, who along with her husband never sold their lands to the MCL, their willingness to accept *gente de afuera* ("outsiders") like the architecture students and MCL employees into the project they initiated derailed their original plans, which were specifically intended, at least initially, to focus on agricultural production: "We made a series of errors, one of which was to allow both the League and the architecture students to come work with us. Our original vision for the project was to make a agricultural project, and to produce beans, corn, milk, all so that we would have food in San Gerardo. The students and the League redirected the project toward ecotourism and biological research. In spite of all that, we wanted to continue with the League, but we needed to do it on our terms."

For ideological reasons—this was a "participatory" project, after all—the MCL also expected the *socios* of San Gerardo Project to take initiative in the implementation of the pilot/master plan, although administrators were often disappointed by what they considered to be a lack of leadership among Project *socios*. The director of the MCL during this period admitted to me that the transfer of lands to the MCL "may have upset the evolution of the social aspects of the project." He did not blame this on the MCL's emphasis on land purchase, however, but on the lack of willingness of San Gerardo residents to take a leadership role in the Project. He told me, "As soon as people saw money, they changed their story—I want my money and I'm gone. So we began poorly when out of twenty families, only around five or six remained. More importantly in this process is that the community's leaders who could have given a push to the project, given it sustenance, sold their lands and left too."

Both sides asserted that, financially and politically, they were equal partners. For MCL administrators and board members, this attitude was partly the consequence of their inability to be fully involved in the San Gerardo Project (even as lower-level employees in environmental education were involved). One Monteverde resident who served on the MCL board of directors during this time explained: "I think of the San Gerardo Project as a footnote in the history of the League. People periodically talked about it. The main administrative concerns were elsewhere, like purchasing lands in other parts of the Children's Eternal Rainforest and ordering all the legal transfers and the paperwork that went with that." During the early 1990s, the MCL was also growing as an institution and adding new employees to programs in reforestation, environmental education, forest patrols, and accounting, and opening a new office and new environmental education programs in La Tigra de San Carlos (a community on the Atlantic side of the MCL's growing forest preserve).

There is a key difference between the involvement of the San Gerardo Project *socios* in the planning and implementation of the Project, however, and that of the MCL employees and architecture students: the latter "participated" in their capacity as paid employees and degree-seeking researchers, while the former worked as unpaid volunteers with the expectation of eventual remuneration. This became particularly problematic as the years passed and still no tangible advances had been made in establishing viable productive activities that could support people actually living in San Gerardo Arriba. By 1992, a year after the publication of the master/pilot plan, only several years after MIRENEM invited the MCL to become involved in San Gerardo, the men and women of San Gerardo Project still did not live there, and their "participation" in the Project essentially meant attending sporadic workshops and trips to improve roads, trails, and ongoing agricultural projects (cattle and pastures, a banana patch, and organic vegetable gardens). In contrast, MCL employees and foreign volunteers attended meetings and engaged in activities in San Gerardo as part of their jobs or short-term commitments to the MCL. According to one MCL employee who worked with the San Gerardo Project, "The fact is, the difficulty that *socios* had of making a living outside the Project undermined what we were trying to do. Those of us who were employees, we were receiving our salaries no matter what we did, so we could sit in meetings all we wanted." Furthermore, participation for *socios* did not simply mean attending meetings: even though there was no formal protocol for sharing expenses, MCL administrators expected *socios* to make financial investments into the Project. So while the MCL would often provide basic resources such as food, construction materials, and transportation to its foreign volunteers, it was often assumed that *socios* made their own investments or arrangements.

This last issue was an especially acute problem for the *socios*, who did not have much disposable income for the project. Their challenge was to secure funding independently of the MCL and its donors, but the legal irregularities of their case (people living within a nationally protected forest reserve) and their ambiguous relationship with the MCL complicated matters. It was a classic catch-22: to provide loans, banks required them to demonstrate that they had legal status as an association, but in order to gain legal status as an association, the *socios* of San Gerardo Project had to prove that they had signed a *convenio* (pact) with the MCL that outlined the formal responsibilities of each group. To avoid legal and financial liabilities, MCL administrators were reluctant to sign any formal agreements until the *socios* formed a legal association. Even though the people of San Gerardo never did successfully form an association, they were able to eventually sign a *convenio* with the MCL in 1992, which would last for three years.[3]

Further complicating an already complex scene is that within the MCL, there has been neither clear consensus on the importance of a widespread and inclusive membership base nor unified expectations of what the "participation" of *afiliados* (affiliated members) entails. As an *asociación* under Costa Rican law, the MCL is composed of *afiliados* who have the right to approve or deny institutional policies in annual assemblies. In its early years, participation within the MCL was defined in terms of being an *afiliado* of the organization and attendance at such assemblies. Work for the organization above and beyond this basic responsibility implied volunteering time and personal resources without any particular expectation of personal gain or remuneration, reflecting the not-for-profit ideology of the North American founders. In the first year of the MCL, the organization was run by its (mainly) North American founders and members, who shared this philosophy and in many cases invested their own financial resources. However, as their organization grew in size and scope of action, the personnel became increasingly professional and Costa Rican. By the time the San Gerardo Project was in the planning stages (1989–1), the MCL's daily operations were already undertaken by a largely Costa Rican professional staff.

During the late 1980s and early 1990s, the MCL's membership was growing and its membership demography was changing. In its first several years (1985–7), when the MCL was largely North American and meetings were held in English, there was general interest in involving Costa Rican residents of Monte Verde as members of the organization, based on the general conviction that the organization could not survive in the long run if Costa Rican residents did not support it. However, there was concern among some of the founders that allowing membership to anybody who wanted to join could undermine the organization's biodiversity preservation mission, since a minority interest group could gain control and redirect institutional policies and resources away from rain forest habitat preservation and rehabilitation (MCL 1988d).[4] To become a voting member of the MCL, a person had (and still has) to prove their dedication to the MCL's mission of "preserving, conserving and rehabilitating tropical ecosystems and their biodiversity" by writing a letter of solicitation, presenting two letters of recommendation from current members, and then undergoing an interview with a panel of members. These requirements tended to discourage Costa Rican *campesinos* from joining the organization, for as one Santa Elena dairy farmer observed, "Everyone around here knows me. Why should I take the time to write a letter and have an interview?"

Nonetheless, by the early 1990s a number of Costa Rican dairy and coffee farmers joined the MCL as a result of their involvement as beneficiaries of the MCL's reforestation program, pushing the overall membership of the organization above one hundred.[5] This process was bitterly contested by some prominent MCL members (primarily some North American and biologist residents of Monteverde), who feared that the new members did not share a similar philosophy of participation and volunteerism based on a lack of expectation of personal material gain. They believed that these people joined to gain access to MCL resources to improve agricultural production on their farms and did not share the commitment to biodiversity conservation and would not work toward those goals. One former member of the board of directors, a North American biologist who resides in Monteverde, explained this position: "In the U.S. people become members of non-profits with the idea of seeing how they can support the mission of the organization. This is a philosophical difference you don't get here, where people join with the idea that they can get something out of it. So associations don't always work well here. Members have a great deal of power, and when you have manipulative people, you have problems."

Indeed, for some of these farmers, gaining access to more resources *was* the reason for becoming members of the MCL, and when the reforestation program dissolved in the mid 1990s, many of these members allowed their membership status to lapse. It is possible to argue that even though they were ambivalent about the MCL, San Gerardo Project *socios* tolerated the ambiguities surrounding their relationship with it in the belief that it could offer them resources for the project they still hoped to create. Indeed, several had joined the MCL as *socios*, and tended to approach their membership in the MCL in the same way they approached their membership in the San Gerardo Project: as dependent upon their own family's needs, and not in terms of what they thought the MCL needed from its membership. Viewed in these terms, it is not surprising that members of the San Gerardo Project had long prioritized building the road to Santa Elena, a school, and agricultural projects over others, since they viewed these as their most immediate needs for being able to make their project work.

In spite of the increasing diversity of MCL membership, the daily operations of the MCL have been undertaken first and foremost by the employees, and secondarily by the board of directors. The employees, while beholden to decisions made in general assembly, nevertheless often respond to a different set of issues than the organization's membership, including the daily concerns of managing the different programs of the organization. These responsibilities often lead to ad hoc actions based on compromises between different actors, including pressure groups within the membership, donor concerns, governmental representatives, and especially the influence of other employees. Furthermore, employees help define the organization's priorities by pursuing relationships with funders, whose interests do not necessarily match the concerns of all of the MCL's membership and target peoples. This will be apparent in the next section, in which I will explain how the MCL became further implicated in the San Gerardo Project at the expense of remaining Project *socios* by choosing to construct a biological station.

## The San Gerardo Biological Station

During 1992, when the San Gerardo Project seemed mired by inaction and there were no significant projects under way that would provide income in the short term for the remaining five *socio* families, MCL officials proposed the construction of a biological station. It was hoped that the station would serve to unite the residents of San Gerardo with a common purpose and to provide them with a viable source of income to continue in the Project. Significantly, it would also open the Bosque Eterno de los Niños (BEN) to visitation by ecotourists and scientific researchers. While land purchases had not necessarily meant the ascendancy of the MCL as *the* central player in the San Gerardo Project (although it made the MCL by far the largest landowner involved in the project and was the reason for the depopulation of San Gerardo Arriba), it was the construction of the biological station that solidified decisive MCL control over the Project and a shift in the ability of the remaining families to define the Project's goals, methods, and priority activities.

MCL officials approached their international donor network and found several donors who were willing to provide financial aid for technical aspects of the project. According to a high-ranking MCL administrator during this period,

> We started knocking on donors' doors and found that some were interested that we were creating this socio-biotic community, but nobody would invest on the social side, only the technical side.... We decided to accept financing to build the station because we felt like we needed to obtain credibility with the residents of San Gerardo, to show that we wanted to get something going to help them.... It was also something that we could make sense of in the League. We had a hard time with the social part, like supporting the community processes. We were forestry people and biologists, so our experience was not working with people, although we tried to through workshops and we learned a lot.

A May 1992 proposal to the Alex C. Walker Educational and Charitable Foundation of California explained that the biological station would promote scientific research in the biologically rich Children's International Rainforest by providing accommodations and facilities for scientific researchers, student groups, and ecotourists. The proposal also emphasized that the station would operate with full community participation, describing it as a "joint initiative" of the MCL and "involve[ing] the neighboring residents in management and operation of the facility" (MCL 1992b). Through The Nature Conservancy, the Walker Foundation donated $50,000 to the MCL for the construction of the station. Barnens Regnskog, the Swedish children's rain forest fundraising organization, also sent money for the construction of a hydroelectric plant. The fact that these funds were donated to the MCL, and not to the families of San Gerardo, reflects the perception among donors that the MCL was better equipped than the people of San Gerardo to realize the technical goals of the project.[6]

*Socios* reacted ambivalently to these donations. On the one hand, several recognized the importance of finally pursuing a tangible project that could provide a much-needed economic opportunity. Indeed, several *socios* helped build the station and donated wood for its construction. On the other hand, reflected one *socio,* this came at a high cost:

> That was when we lost our autonomy as a group. We lost the spirit of the project, in which the community was supposed to be the central focus. We had planned for a biological station in the project, but it was going to be one of the last things we did, after we had established ourselves economically as a community. It had been our vision that we would develop an agricultural base first, and then we could bring in tourists. You see, we had to be able to feed ourselves before we could feed tourists. But the station became the center of the project.

Even with positive intentions to resurrect what they perceived to be an internally divided and lackluster San Gerardo Project, the decision to channel funds into a biological station had the effect of recasting the goals of the San Gerardo Project in terms of running a facility to provide accommodations for visitors to the International Children's Rainforest. It also further insinuated the MCL in the daily administration of the Project in ways that neither MCL administrators nor *socios* had necessarily expected because it redefined the MCL as the owner-administrator of the biological station, and more importantly, it redefined the *socios* as its labor.[7]

Since the MCL was responsible to its donors for the successful construction of the station and the hydroelectric plant, institutional priorities lay with the elaboration of these particular efforts.[8] MCL institutional priorities now dominated in two crucial respects: the siting of the station and the naming of a coordinator for the station. In the original planning process that produced the pilot/master plan, the architecture students, MCL administrators, and project members had defined the lands nearest to the Santa Elena road to be the area where organic agriculture projects would be developed, and that biological research should be conducted and a station built in a more remote part of San Gerardo. When it came time to actually construct the building, administrators viewed this as a technical process and invited Monte Verde resident biologists to help identify the site of the station.[9] Biologists worried that visitor activity would jeopardize the forests in the remote areas of San Gerardo, and argued that it be placed in already deforested uplands near the road. Furthermore, since the MCL would be entering an increasingly competitive national market of biological stations that catered to scientific researchers, student groups, and ecotourists, members of the committee asserted that relative accessibility to the Santa Elena road would be an important factor in attracting clients. As a result, the station was built in the area defined originally as the site for the agriculture projects, near the road.[10]

A second point where administrative priorities dominated was in naming the coordinator for the San Gerardo Project in 1993. In addition to serving as the liaison between the people of San Gerardo and the MCL, this person was in charge of supervising the building of the biological station. MCL officials viewed the position as a technical one, and they appointed a North American man who visited Monte Verde as a tourist and offered his carpentry skills. Although this individual had a reputation of openness to the people of San Gerardo, and worked very closely in particular with the two ex-MCL forest guards who joined the San Gerardo Project, his lack of fluency in Spanish prevented effective communication with the *socios,* none of whom spoke English. At least one member of the San Gerardo Project partially blamed his leaving the project in 1993 on his inability to communicate with this man and the fact that the MCL would hire an *extranjero* (foreigner) who was not from San Gerardo in the first place to coordinate the establishment of the biological station in what had been designed and advertised as a community participation project.

Another reason this man cited for leaving the project at this point was that he felt the terms of his *apoyo* for and *inversión* had changed: he now felt like an employee of the MCL, instead of an equal and independent partner. He remembered bitterly,

> Even before the station was finished, the MCL started bringing student groups and donors to the biological station. They would introduce us to the people as the "socio-biotic community." It's like we were little rabbits they would bring to show the donors. But when it came to taking meals at the biological station, they didn't want us to be eating there at the same time as the visitors and told us to hide the children. They told us that there was not enough room. They also did not like the fact that we had cattle there. I realized that the MCL administrators who complained of these things didn't think of us as equals, but as employees they could order around.

While these actions by themselves do not necessarily constitute singular reasons for the dissolution of the San Gerardo Project, two families cite them for resigning from the Project in 1993 and 1994, which further weakened their efforts to keep the Project going. By 1994–5, the number of *socios* of San Gerardo Project was reduced to three families, the two ex-MCL forest guards and one of the families that had not sold its land to the MCL.

## Conclusion

In an annual report sent to donors during the mid 1990s, the MCL's San Gerardo Project coordinator admitted that the number of *socios* was lower than desired, but expressed optimism that the few who were left were actually now living part-time in San Gerardo (while they built the station) and that the socio-biotic community could be resurrected if housing were built and several of the families that had left were invited back (MCL 1994b). This latter concern, that of housing, was never taken seriously by MCL administrators, who never undertook fundraising for the purpose. The very idea of facilitating *more* urbanization in a fragile ecosystem would no doubt have generated severe conflict among the membership.

In early 1995, the MCL changed its director, bringing a new phase in the San Gerardo Project: its formal dissolution. The new director sought to bring the San Gerardo Project under more direct administrative control and make it less of an independent fiefdom. As a result of this process, a debate emerged among employees and MCL board members about whether or not a "community" existed any longer in San Gerardo Arriba, and whether what remained could be revitalized in terms of the original plan for the socio-biotic community project.

Reflecting on his first visit to San Gerardo Project in 1995, the new MCL director expressed his disappointment:

> When I came to the MCL everyone was talking about San Gerardo, about the socio-biotic community, and I thought that for the first time, I might really experience a sustainable development project. When I arrived at the biological station, I thought, gee, this is strange, there doesn't seem to be a community here, from a physical point of view. From the point of view of human relations, what I saw was the coordinator [the North American carpenter] and what looked like a couple of employees of his [the ex-guards] and some North American volunteers working on the station and organic garden.... I thought that this was the biggest hypocrisy I'd seen, because one would read these documents and hear talk about this community, but this was nothing like what they said it was supposed to be.

The new director urged that the Project had to be legally dissolved—the pact between the original five families and the MCL declared defunct—since the original goals of the San Gerardo Project were unmet and only three families remained. It was a difficult process for some employees and board members, for as one employee who worked closely with the Project throughout its history told me, "Some of us had a strong emotional attachment to this Project and the ideals it represented. We really thought this could be a model for sustainable development, since it would allow people to continue living in the forest instead of removing them." In late August 1995, the men of the remaining three families met with the MCL board of directors at the San Gerardo biological station to define the future of the project. They agreed to dissolve the pact they had signed with the MCL, in exchange for compensation for their labor and material investment into the biological station.

Among ex- *socios* of the San Gerardo Project, there is profound ambivalence about the meanings and prospects for "sustainable development" in the Monte Verde region. Several years after she left the project, one woman reflected cynically on her experience:

> I think the San Gerardo Project could have been sustainable development if it were only allowed to proceed over more years. It could have been a model even. The ironic part is that the League got involved and redefined it. Maybe because of a technical mindset they weren't looking at us, the people who lived there. We had convinced ourselves it could work so we didn't necessarily see these problems until late. The scheme we had could have worked—could still work—but the League broke the scheme by pushing us into tourism and the biological station.

This comment reflects a key insight about San Gerardo Project, which is that at a crucial point in its elaboration, the people of San Gerardo, who believed themselves to be subjects of their own social and political process, came to understand that their priorities were redefined by the involvement of the NGO. It was seen to be the result of a fundamental inequity between the NGO (with its access to knowledge and international funding) and the people it sought to help. But inequity describes only part of the dynamic, emphasizing the negative role of the NGO over the inability of Project *socios* to reach consensus on their uneven commitments to the Project in the first place. Furthermore, it does not demonstrate any awareness of the varied perspectives within the MCL itself on the San Gerardo Project and the participation of Costa Rican *campesinos* within organizational processes. It is here, on these contested and negotiated grounds where participation (or *apoyo*, or *inversión*, and so on) is defined in practice, as the result of culturally specific attitudes toward processes of social and economic development. It is also here that institutional processes have effects beyond what actors anticipate, and can result in a contradictory situation in which the priorities of institutional growth and maintenance eclipse the multiple concerns of its members and target peoples.

MCL officials never considered their work in San Gerardo to be "governmental," if that word is meant to describe a centralized body that provides services or coordinates infrastructure processes. They regularly emphasized that the *socios* of the Project were "equal partners," if they did not necessarily clearly communicate this to them. But there were definitely elements of governmentality that it came to assume, in the absence of consistent unity among Project participants and any other external agency, such as the Costa Rican state, that might have offered a counterweight to its involvement. The crucial turning point was when it began purchasing and concentrating its land holdings in San Gerardo, and thereafter—albeit in fits and starts—began centralizing the goings-on of the Project out of its offices and with its employees. When it undertook fundraising for the biological station and then built it, it asserted a vision of San Gerardo as a site for scientific research and touristic visitation, in the process redefining *socios* as paid employees and the rest of the Project's initiatives—organic farming, road maintenance, and the like—to be in service to those needs. While some of the remaining *socios* actively expressed their displeasure at these turns of events and believed their project was highjacked and their good names being abused by the MCL in its fundraising efforts, they nevertheless lacked power to redefine the the Project's momentum once it was on this track.

Given the inclusion of San Gerardo lands in the Arenal Watershed back in the mid 1970s, the question arises: would not San Gerardo as a community have disappeared anyway? Critics might say that in the face of sustainability initiatives and state-required landscape protection, the *campesino* way of life as practiced in San Gerardo, based on expansive modes of agriculture, is outmoded, if not downright inappropriate. But such an assertion assumes a basic unwillingness or inability on the part of rural Costa Ricans to adapt to changing circumstances, while masking the structural inequalities that exist between such communities and the NGOs who claim to help them. In fact, turning the lens on the NGO itself reveals that as the neoliberal state increasingly relies on them to carry out certain kinds of public sector initiatives, questions of accountability rise to the surface. It is apparent that MCL officials were accountable to their organization's donors and members (some members, perhaps, more than others), but the question of their organization's accountability to the *socios* of the San Gerardo Project was never really settled. The dismemberment, that is un-membering, of the San Gerardo Project was by no means inevitable. But given the territory of complicated legal arrangements in which one side was able to gain access to funding and the other was not even able to formally create a legal association, as well as the ambiguous and contested definitions of participation on both sides, the San Gerardo Project's chances of long-term success were up in the air. To take one key element of that dynamic—the NGO itself—out of the equation, and explain the project's dissolution on the basis of the ignorance or inabilities of *campesinos* to make it work, is at best unfair, and at worst, perpetuates an uncritical acceptance of the role that NGOs are increasingly playing in the politics and everyday realities of public sector governance in neoliberal Costa Rica.

A final question arises, which is why the San Gerardo story is dismembered from the tales of Monte Verde conservation successes, that is, swept under the rug of institutional memory in the MCL and excluded from the mainstream tales told to tourists and journalists who come to visit. One central reason, perhaps, is that within the MCL there has been a massive turnover in personnel, bringing new people into the organization who did not live through this experience. Even more importantly, it has also been painful and disenchanting for those people— *socios* and MCL employees alike, though for different reasons—who invested their hearts, souls, and material wealth in the project, to see it dissolve in the end. Keeping the lens on the MCL, we could ask, as Chambers (1993) does, whether or not the MCL was a "learning organization," that is, able to adapt to changing circumstances. It is difficult to say, partly because there is no stable or singular "it" to refer to here, although "it" has certainly learned that many *campesinos* not involved in the San Gerardo Project have regarded the organization with deep skepticism as a result of its role in the dismemberment of San Gerardo. One problem with Chamber's formulation, that organizations can be learners, is that besides the fact that staff and board members rotate in and out, compromising institutional memory as it were, there are often such profound philosophical divisions among members at crucial moments in institutional history—as we have seen here in regard to "participation" and "sustainable development"—that institutions are not always in a position to overcome internal divisions to learn. Further, it is important to explore how external processes or events beyond the control of the institution may undermine cases in which key members or staff people learned. It would seem to be the case in San Gerardo, where one key reason that contributed to a shift in project priorities was the debt-for-nature swap that started the land sales back in the early 1990s. Previous to the swap, it seemed as if key MCL staff and members had "learned" that land purchases were not necessarily the best mechanism for conservation; yet once the money came, they had little choice but to spend it for what it was intended to do, that is, create the conditions to move people out of San Gerardo.

# Notes

1. Although there has been a tendency to collapse "sustainable development" and "ecodevelopment" in recent years, the latter has insisted on highly specific solutions to particular problems in specific cultural and ecological contexts, while the former has implications of national level capitalistic growth (Romero ibid.). Some of these tensions played out in the elaboration of Costa Rica's Estrategia de Conservación para el Desarrollo Sostenible (ECODES; Strategy of Conservation for Sustainable Development), that has been criticized for its dismissal of the anti-statist rhetoric of "ecodevelopment" (Mora ibid.; Quesada and Solís 1988).

2. Commercial debt-for-nature swaps were developed in the late 1980s and early 1990s by the WWF-U.S., Conservation International, and the Nature Conservancy, and as of the mid 1990s generated at least $109 million for conservation activities worldwide. In a debt-for-nature swap environmental groups purchase a "developing" country's foreign commercial debt at a discounted value in the secondary debt market, and subsequently exchange the debt in return for the obligation of the debtor nation to provide favorable conditions and financial support for environmental conservation programs. Between 1987 and early 1991, debt-for-nature swaps generated about $42 million for conservation in Costa Rica at a cost of $12 million (largely in donations from the Costa Rican government and northern environmental activists and organizations). This mechanism reduced Costa Rica's national foreign commercial debt by $79 million (Quesada Mateo 1992b; Williams 1992). It is argued that in a debt-for-nature swap, all parties win, since banks that have loaned to developing countries receive at least a portion of their money back; the countries reduce their debt inexpensively; and environmental organizations gain funds to carry out land purchases, for endowments, and for environmental education programs. In Costa Rica, debt-for-nature swaps were coordinated by MIRENEM and the Banco Central, along with the WWF, Rainforest Alliance, and The Nature Conservancy. Although MIRENEM received the major portion of debt-for-nature swaps, NGOs like the MCL played a central role since organizers and sponsors of swaps widely considered that NGOs knew the 'on-the-ground' realities and needs of communities.

   Although sponsors and organizers of debt-for-nature swaps have been careful to note that they cannot require the beneficiary to spend the money as the sponsor wants, there is no doubt that continued access to such monies depends on spending the money in the ways agreed upon between sponsors and beneficiaries. The MCL's credibility with the sponsors and international conservation donors would have certainly been questioned, and future access to funds reduced.

3. *Socios* wanted a long-term *convenio* that would last twenty years, believing that they needed a long time to get the project established. MCL administrators wanted a much shorter *convenio* (two years) as a probationary period to see if the project could really work. In the end, the *convenio* was negotiated to last for three years.

4. MCL members with this perspective invoked the case of ASCONA (Asociación para la Conservación de la Naturaleza), the first environmental NGO in Costa Rica, whose membership base shifted in the late 1970s as politically motivated opponents of ASCONA joined as members to undermine the organization from within. The risk that the MCL be taken over by hostile interests continues to concern certain MCL board members and members, who argue that the MCL remains vulnerable to a "fox in the henhouse" situation in which a particular interest group could wrest control of the administration from the current leaders and liquidate the assets of the organization for personal profit.

5. This number has been around 150 for the past decade.

6. Although this does not mean that relations with donors have always been free of tensions. For example, relations with the Walker Foundation, while largely positive, have had their tensions, such as when the Walker Foundation requested that the biological station be named for a person of its choice. MCL officials resisted this on the grounds that this was an inappropriate idea

for a community project like San Gerardo Project. Relations with the Swedes have also been marked by tensions; they have been dissatisfied with the progress of the construction of the hydroelectric plant.

7. The MCL paid *socios* 150 colones (approximately $.85) per hour for their help on the station. This wage was below average for such work, but MCL administrators expected the *socios* to contribute one-quarter the value of their labor, to demonstrate their commitment to the Project.

8. It is important to point out that these were not necessarily the priorities of individual employees of the MCL who worked on the San Gerardo Project. For example, several MCL employees who had close contact with the San Gerardo Project, including a sub-director who was involved in the project since its early stages and a coordinator of the project who entered later, argued among themselves and in annual reports that much more attention should be paid to maintaining a viable community of people, and that the establishment of a tourist industry in San Gerardo should be secondary (MCL 1994b).

9. The MCL director commented that involving biologists in this process was an important political move: "This process [siting the station] involved the community, technical people in the League, and the biological community in Monte Verde. The biological community had been very negative toward the League during this time, since we had been making decisions without them, and well, you know the problems when you don't take the biologists into account." *Socios* of San Gerardo Project have disputed this version, denying that they were ever invited to be part of this process, and maintaining that it was mainly biologists and MCL employees who worked on this process. In fact, out of the five families, only two of the men were part of this process.

10. Critics of the biological station, including several *socios* of the San Gerardo Project, have noted that the station site the MCL chose also has a striking view of Arenal Volcano and Arenal Lake. The original site proposed for the biological station did not have these attractions. They argue that the MCL was more interested in the station's potential to bring tourists than it was in following the plans of the people of San Gerardo. It was confirmed in the minds of some when the MCL accepted donations from the Grand Circle Foundation, a branch of Grand Circle Travel that is known for the ecotours it organizes for North American senior citizens. By 1994, Grand Circle had donated $80,000 to the MCL to support the Bosque Eterno de los Niños project.

# 6 Contesting "Community" in a Community Conservation Project: The Fight for the Reserva Santa Elena

## Introduction: "Pueblo Pequeño, Infierno Grande"

Soon after I arrived in Santa Elena during 1995 for a year and a half of fieldwork, I met with a friend who had been working at the Reserva Santa Elena (RSE), to catch up on what was happening there. He eagerly launched into a long and detailed description of what he called "*pueblo pequeño, infierno grande,*" or "small town, big hell." It is a phrase one often hears in rural Costa Rica to refer to the way in which seemingly slight affronts or tensions can explode into widespread conflicts.

The bare outlines of the story he told me were the following. A group of *caciques* (chieftains or community-level leaders who tend to dominate in formal political processes), calling itself the *Fundación* ("Foundation"), had taken over the RSE and had been quietly running it for several years. Administrators and parents from the Colegio Técnico-Agropecuario de Santa Elena (Santa Elena Technical-Agricultural High School), who asserted that the Colegio was legally entitled to operate the RSE, had been pressuring the Fundación for an accounting of its operations, and for it to relinquish control over the reserve. Tensions were building in Santa Elena, and there was talk of a popular demonstration thirty-five kilometers down the mountain on the Panamerican Highway, to bring wider attention to this crisis. At the same time, I had noticed flyers popping up around Santa Elena village at the *pulperías* (grocery stores) and other spaces where public announcements are generally made. One of them declared in handwritten Spanish:

People of Santa Elena!

The Colegio's Reserva has been invaded by the Fundación in an illegal and offensive manner to our Colegio. WE CANNOT PERMIT THIS. We need your solidarity. Denounce them yourselves. Communicate this abuse of the Colegio and Community with the press, radio, t.v. The Colegio is open for your information.

Thank You. The People.

During the next few days, I learned very little else about the situation (except that representatives of the Fundación skipped the meeting). People seemed reluctant to speak about it with a newcomer like me. Then one morning a week or so later my neighbor Don Roberto, a member of the Colegio's five-person Junta Administrativa (school board), came over to my house and asked me if I wanted a ride to a demonstration at the RSE.

That morning, the tensions that had been slowly building for the previous nine months or so came to a dramatic head when about fifty people convened at the entrance to the RSE to protest against the Fundación (see figure 6.1). The crowd was largely made up of students from the Colegio (who hail from small villages and hamlets dispersed around the region), accompanied by a handful of teachers, administrators, parents, members of the Colegio's Junta Administrativa, alumni, and representatives of the Ministry of Public Education. The intended target of this demonstration, Juan Montoya, the man who was the RSE director at the time and one of the Fundación's founders, stood by poker-faced. The goal was simple: since negotiations had failed, the Colegio was here, if not to take the RSE back through direct action, then to at least show how serious people were about taking it back under their control.

A little after 8 a.m., two solemn students brought the mingling crowd together when they marched across the porch, one with a Costa Rican flag and the other with the Colegio's banner, and led everyone in the national anthem. The speeches began with a rousing one about the community's right to control this property by the Colegio's principal. The speakers who followed—local leaders, school alumni, a representative of the Ministry of Public Education—railed against the Fundación in passionate voices, claiming defiance and determination. As new speakers got up, cheers and applause became more enthusiastic, and Don Roberto turned to me and ordered me to clap harder. A young alumnus of the school, who within six months of this date would be the director of the RSE, spoke impassioned from the crowd, to which a man cried out "¡Así es hombre!" ("That's it man!"). A chant began: "Give us the keys! Open the doors! Give us the keys! Open the doors!"

*6.1. Students on their way to the Reserva Santa Elena demonstration
(photo by author).*

After some vigorous chanting but no clear submission to their demands, the demonstration petered out: people broke into small groups to talk, wander off to the parking lot, explore the edge of the forest, or talk to the policeman about the latest gossip, and some simply went back to Santa Elena. Meanwhile, members of the Colegio's Junta Administrativa and the Colegio principal approached RSE director Juan Montoya and once again sought to negotiate with him.

In the following days, rumors began circulating about threats of violence and the potential for mob justice coming from both sides of the conflict. One friend who was close to the Junta Administrativa of the Colegio during this period told me,

If it weren't so tense, it would have been almost funny. That afternoon after everybody had gone home from the demonstration, some representatives of the Colegio went to make an inventory of the Interpretation Center, with a lawyer. All of a sudden all these guys who had sided with the Fundación, some of them were drunk, right, showed up and were insulting everybody and making allegations. Well, they closed up and left and everyone thought that was it. But that night a couple of men who were on the Junta Administrativa of the Colegio stayed there to protect the place, I mean we thought it was ours at this point. The next morning we got there at like 7 a.m., and there these two of ours were, out in the parking lot, and the Fundación people were inside! [Laughs.] I guess these Fundación people had gotten there at 5 and broken the door down, and wanted to grab the two who were inside. So the two inside came out, one with a machete, one with a pistol, while the others surrounded the place. After a while the two, outnumbered, decided to leave. When word of all this got down to Santa Elena, people were livid, and one guy got his gun and said he was going to shoot the sons of bitches in the Fundación. He shot his gun in the air a few times. I think there were some people disposed to fight with machetes that day. We had to calm ourselves down, and fortunately nothing ever happened because we realized things had gone too far. So we decided the legal and public relations routes were better to take. This was the first time we've ever had a confrontation this severe here in Santa Elena.

He added that "up the hill in the gringo village of Monteverde," there was concern that this conflict could jeopardize the conservation of the forest. In response to this he said: "This was never in question, this was about other issues that had been building for some time."

## The Contours of a "Community" Project

This was a striking situation that immediately complicated my understanding of the RSE and its status as a community conservation initiative. Since its founding in the early 1990s, the 310-hectare Reserva Santa Elena (Santa Elena Cloud Forest Reserve, or RSE) has been as a self-consciously "local" response to the widespread (if somewhat simplistic) perception that Monte Verde's (and Costa Rica's) natural resources and tourism economy have been dominated by outsiders and foreigners, such as the San José-based TSC or MCL. It began when Santa Elena's chronically underfunded technical-vocational high school (*Colegio*) was able to secure a concession from the owners of the cloud forest, the government Ministry of Environment, Energy and Mines (MIRENEM, whose name has since been changed to MINAE, Ministry of Environment and Energy), to operate an "ecotourism and research project" for the financial and educational benefit of regional public schools. Advocates emphasize the uniqueness of this situation: "Santa Elena Reserve was created out of a community's determination to share more of tourism's benefits, and to use its resources as a tool for direct community development.... The Santa Elena Reserve is one of the first community-owned and community-administered reserves in the country. It is an excellent example of what ordinary people can do to conserve their environment" (RSE 1991). At some twenty thousand visitors a year (in 2004) who each pay $8 to enter, the amount of money the reserve generates is substantial.

Unlike the San Gerardo Project, which has been largely erased from Monte Verde's catalogue of successes and associated with a "private" membership-based association (the MCL), the RSE has been highlighted as one of the region's most positive examples of home-grown community involvement in conservation, managed through the public institution of the Colegio. Every year for the past five or so, the RSE has sponsored a festival to celebrate its status as Monte Verde's main community-based attraction, and Santa Elenans turn out in droves to affirm that. This image has been picked up and circulated by the government's ICT (tourism promotion board), virtually every tourism guidebook on Costa Rica, and transnational ecotourism scholars who tend to celebrate it as an example of "genuine ecotourism" (Wearing 1992; Wearing and Neil 1999; Honey 1999). One of these scholars observes, "This project seeks to achieve sustainable development and fulfill the development requirements of the population surrounding the project. This is in the belief that it is only when conservation projects benefit local communities, and are set up with an infrastructure that vests control within local communities, that genuine ecotourism is achieved" (Wearing 1992: 125).

This chapter explores how the RSE became identified as, and what it means to be, a "community" cloud forest preserve and tourism attraction controlled through a public institution. My argument is that even while the imagery of local community involvement and benefit is *the* defining feature of the RSE, and in spite of such inclusive convictions, there is no singular or even stable concept of community operating there, if community is thought of as a collective of shared values and social, geographical, political, and economic consistency, homogeneity, or unity. In fact, there are meaningful contours and contradictions in how Santa Elenans and their nonlocal collaborators have defined and redefined this place as a community project, which reflect both fissures and alliances based on factors like political philosophy, economic status, access to financial resources, kinship, and friendship, among others. Since even before its invention as a conservation and ecotourism project, what would become the RSE—as well as the RSE itself—have represented a discordant arena in which national and regional economic conditions have been refracted, the Costa Rican state's shift toward neoliberal public policies has been actively supported and contested, and Santa Elenans' own attitudes toward self-determination and rapid social change have been considered and debated. Furthermore, its connection to the high school has varied, and while now quite firm, it has been quite weak.

As such, the RSE's identity as a "community project" can obscure more than it reveals, that is, the highly differentiated, contingent, and flexible ways people's identities as community members have been defined through the project, and how these processes reflect and derive from ongoing struggles over the conditions of life and conviviality in Santa Elena. It is true even today, when the conflict I describe here has long since calmed down.

In the conflict described here, the issue was not simply who has control over the administration of the RSE, but the very meanings and practices of membership in the moral and political life of a rapidly changing Santa Elena. It was rumored that one faction in the conflict, the Fundación, had intended to use their government connections to push for the semiprivatization of the high school, funding a new school and hiring personnel for it with the profits made on tourism at the RSE. School officials and parents of students, many of them relatively poor *campesinos* living in the hinterlands outside of Santa Elena, quickly organized themselves to resist such a move, asserting that privatization would benefit the children of elites at the expense of free public education for all. While the one side emphasized the importance of private administrative experts controlling community resources for the benefit of many, many on the other side envisioned the management of collective resources through a public institution. At its heart, this was as much a struggle over class tensions and the shape and funding of public education as it was about who should manage the ecosystems and tourists at the RSE. Although most admit that it was unusual in its intensity, the conflict demonstrated (or better, reinforced) for many residents that their desire for and ideals of local community control are just as much concepts and relationships that are hotly contested, to be bought, sold, normalized, or imposed, as they are to be admired.

As an icon of community-based sustainable development, the case of the RSE poses interesting complexities for natural resource and ecotourism planners and practitioners who have been recently affirming the significance of community involvement and benefit in conservation and tourism projects. There is a strong functionalist bias in environmentalist representations, where "community" tends to be identified as a group of people with common values, practices, and beliefs, with an emphasis on their ties to a specific geographic and natural place (Murphy 1985; Adams 1990; Boo 1990; Wells et al. 1990; Whelan 1991; Adams and McShane 1992). Barham (2001: 185), for example, defines community as "a sense of attachment to the land and a relation to it that entails responsibility and intimate knowledge." Seductive in its firm associations of people and place, such a definition is nevertheless based on a unidimensional definition of people's lives, seeing community more or less solely through ties to a specific physical landscape, especially at a moment in which Latin Americanist anthropology has been rejecting the model of the "closed corporate community" (Wolf 1955; Rouse 1991; Kearney 1996; García Canclini 1999) in favor of paradigms of disjuncture, mobility, and contestation.

More importantly, the very geographic or territorial space entailed in RSE definitions of community is neither stable nor bounded. Since its invention, the RSE has been what Hannerz (1990) calls a "cosmopolitan" organization, that is, an organization in which social relationships between participants are mediated without logical necessity to particular areas in physical space, nor are participants easily defined as "locals" whose interactions happen only in face-to-face contexts (Peters 1996). Even while rhetorically reaffirming the RSE as a community project, participants have drawn strategically on international and national institutions and allies in order to bolster their claims of representing their community, as well as to resolve the conflict described in this chapter. Given the fractious and unstable character of these processes, whose version and what territorial vision of community do planners and allies normalize when they offer their solidarity and assistance? As I will show here, participants regularly manipulate community discourse to normalize what and who they include in their actions, and nonresidents have both consciously and unconsciously contributed to the normalization of, and in some cases reoriented, those definitions.

## "Of Costa Ricans, for Costa Ricans": Reserva Santa Elena as "Community" Initiative

Since at least the late 1940s residents of the Monte Verde region have considered the land that is now the RSE marginally productive for agriculture because of its wet and cold climate, poor soils, and remoteness from Santa Elena or any other villages. Unlike farms in the Monte Verde region at lower altitudes, the Finca Río Negro (its original name) therefore never experienced massive conversion of forest. By the late 1970s only 17 percent of the land's forest had been converted to pastures for high quality Angus beef cattle. This venture failed, however, due to the inability of the cattle to withstand the cold temperatures. This area, whose average elevation lies at 1,600 meters, is commonly known among *campesinos* as "the Refrigerator," and legends refer to frozen cattle and people almost dying from the cold.

Because it fell within the Arenal dam's protected zone, the Finca was expropriated in 1979, and ownership passed to the Ministry of Agriculture and Livestock (MAG). At the same time, the Colegio in Santa Elena was about to graduate its first students with degrees in agriculture and animal husbandry. Because of downturns in the agricultural economy and lack of affordable productive land in the area, the Colegio proposed a "parcelization project" at the Finca Río Negro that would provide young graduates with productive opportunities. The project was approved in 1982 and the high school gained a ten-year concession on the farm, to divide it into ten-hectare parcels to be assigned to graduates to work as an agricultural cooperative (CTASE 1982: 3). After several years of arduous work but failure to scratch out any sort of productive existence due to the farm's distance from town, the muddy trail (for at this point a road had not yet been created), and harsh climatic conditions, the recent graduates began to give up and look elsewhere for jobs (which for many meant finding a place in the emerging tourism industry as bartenders, waiters, and hotel receptionists, or migrating to find work elsewhere).

Interestingly, in language around the parcelization project there is no reference to "community" control and benefit that would become so prominent less than a decade later in the proposal to open an ecotourism and conservation venture.

In fact, the parcelization project proposal dwells on parcelization as a solution to a "national" crisis in agricultural production, and of the "grand national interest to incorporate these young people in the productive process, a necessity that is accentuated in these moments of crisis that makes national [agricultural] production indispensable" (ibid.). According to the proposal, the main problem for these graduates was access to land: "Many have not been able to involve themselves in national production because they do not have access to the necessary resource, land" (ibid.).

This project was mainly managed and financed by Santa Elenans themselves, so why did it emphasize national, not local themes? In the late 1970s, economic trends and Central American regional political crises led to what is commonly called the "debt crisis," which was playing out in the early 1980s through structural adjustment programs and neoliberal reforms (Edelman and Kenen 1989; Rovira Mas 1989). In Monte Verde, where dairy farming had expanded throughout the 1960s and 1970s to become the region's primary economic activity, production decreased and some farmers faced with trying to make ends meet began to look into alternatives beyond dairying. The Colegio's board of directors, many of whom were themselves dairy farmers, were concerned that the nationwide crisis in the dairy industry would prevent their first graduates from finding work, a poor precedent for generating support for the new school. MAG approved the project, since as the government ministry responsible for the land, it was not eager to spend its own limited resources keeping invaders out of its farm, and furthermore it would receive a small rental fee for use of the land.

With the failures of the parcelization project, work at the Finca Río Negro went more or less dormant, although there were various attempts by squatters to occupy the lands. By the late 1980s, the idea of creating an ecotourism reserve began circulating in Santa Elena, due in part to the fact that area hotel owners were pushing to create more tourism attractions. During the 1980s increasing numbers of tourists began to seek out Monte Verde, attracted by the Monteverde Cloud Forest Preserve. The most dramatic increases in visitation to the Preserve took place between 1985 and 1989, in which visitation grew roughly 25 percent each year (Aylward et al. ibid.: 327). By the late 1980s, the Monteverde Cloud Forest Preserve administration began to discuss limiting the maximum number of visitors at any one time in the park to reduce pressure on its resources. Although limits were not implemented until 1992, hotel owners recognized an opportunity in opening the forested Finca Río Negro as an alternative trail system to the Monteverde Preserve. Their support for the scheme, both individually and through the newly created Cámara de Turismo de Monte Verde (Monte Verde Tourism Chamber), would be crucial in helping it gain wider regional and governmental acceptance.

Until the initial tourism boom, the centers of formal political and economic power in the Monte Verde region were centered in the Quakers' Cheese Factory and Quaker-centered Town Meeting, where decisions regarding infrastructure were made and actions coordinated. The tourism boom not only offered a quick economic rise for those who were able to build hotels, but it also created a new political presence in the region that still exists alongside, and sometimes in tension with, the goals of the Cheese Factory and Town Meeting. The former owner of the Finca Río Negro, Romero Hernandez, himself became one of the prominent players in this sector. In the lobby of the hotel he opened in 1991, he explained to me the reason he was so invested in helping establish the RSE:

> During the late 1980s, I remember thinking to myself, 'What am I going to do?' If I made a hotel, where was I going to take my guests? Because they weren't necessarily going to let them enter at the Monteverde Preserve. Touristic activity was just beginning to get going, and foreigners controlled it here. I began to write the project, and some of the gringos from Monteverde, they told me a project there would never work, that I was wasting my time. So I told them that this was a project *of Costa Ricans, for Costa Ricans* [he raises his voice, emphazing each syllable].

Hernandez helped organize a public demonstration in favor of the RSE, and engaged in some back-room politics in San José at MIRENEM (to which ownership of the farm passed during the 1980s). The invocation of a mild nationalism in his rhetoric is striking, and indicates how importantly the initial efforts to establish the RSE were influenced by a widespread reaction during the height of conservationist land purchases in the late 1980s against what many Santa Elenans felt was the selling of their national territory to foreign interests.

## International Help for a "Local" Project

Why does the language of locality take over so strongly by the second half of the decade, in fact, within only several years of the nationally oriented parcelization project? How did this language become incorporated into language surrounding the project and why was it important? One reason could be the rise of community-based ideologies in transnational sustainable development discourse. To an important degree it is, as dominant languages of globalization suggest, the result of the "flow" or "diffusion" of the concept from metropolitan institutions, intellectual centers, or traveling cultural discourses (Appadurai 1996; Rosaldo and Inda 2002). But to leave it at that is relatively meaningless, for the significance of the localizing discourses was in its vernacular usage, which in this case attached it to certain public education institutions centered in and around Santa Elena.

The original RSE proposal identified the community participants and beneficiaries as five public education institutions in the Monte Verde region. According to the proposal, profits would be redistributed in the following proportions: 50 percent to the Colegio, and 10 percent each to four primary schools, and the final 10 percent to be reinvested in RSE infrastructure and improvements (RSE 1990). One Colegio teacher explained that 1990 was a particularly difficult year for the Colegio in Santa Elena; enrollment had dropped to eighty students, and the Ministry of Education considered closing the school. The RSE's connection to the Colegio must be understood in such a context:

> We lived a particularly difficult situation. We saw teachers who would leave every Friday afternoon to go to their homes in San José for the weekend. There has always been instability and indifference in personnel here. But that year the principal really turned things around and inspired us when we heard of the possibility that they would close us because of the lack of students, and people began to talk about what should be done about the Colegio. And before that there had been these difficult discussions about the Finca Río Negro and what to do with it, and one of the priorities that emerged in all this was to incorporate the concept of an ecotourism reserve into the sustenance of the Colegio. In other words, it motivated people to work to make the Reserva provide sustenance to the Colegio.

Sufficient funding has been a perennial concern for the Colegio. For example, in the mid 1990s when the fight over the RSE took place, it received only three hundred and fifty thousand *colones* from the Ministry of Public Education for its annual budget (not including salaries). That is, it received approximately $1,750 to cover classroom materials and supplies, maintenance and improvement of infrastructure, and general school activities for over 120 students. The rest of the required money is generated through various school projects, including a dairy operation, a small coffee plantation, and various smaller projects such as raising chickens and pigs, all of which are integrated into the academic schedule and curriculum. The RSE represented a possibility to move beyond the raffles and dances that normally, if inconsistently, offered other financial support for school programs. The redistributive structure of the RSE sought to normalize a certain vision of community through the benefits generated for students, first and foremost of the Colegio, and secondarily, in several primary schools.

It is ironic that an *individual* such as Hernandez has taken credit for establishing a purportedly *communal* project like the RSE. Yet this irony speaks to an interesting situation: that key individuals and institutions, some of them not based in the Monte Verde region, have pushed the RSE's development. In fact, Hernandez's version of his own involvement is contested, and some residents give credit to others, including Juan Montoya, the Fundación's founder, and his connections to international volunteer groups, for being crucial in its development. Reflecting on the project he is credited by some with building, Montoya asserted that "the community" never really helped him at all with the creation of the RSE:

A few of us thought we could develop this as a community project in which community people would participate and benefit. But what we found was that there was little community interaction in this process, and little support from the Colegio. A few of the graduates of the Colegio and a couple of teachers collaborated with me, but we were constantly struggling to convince people of our plans. That's when Youth Challenge International contacted us and said they would help. Without them, it wouldn't have been possible to do it.

Indeed, the early years of the RSE (1991–94) do not reflect any simple sense of community control or involvement.[1] Not only was the "master plan" for the project designed by an American graduate student at Duke University (Kuzmier 1992), but Youth Challenge International (YCI) has been credited with being "the pioneer organization in the RSE project" (Fundación, n.d.: 2). Formed in 1989 and based in Canada, with affiliated groups in Australia, Costa Rica, and Guayana, YCI facilitates international volunteer work for 18– to 25–year-olds on conservation projects like the RSE, as well as projects such as building schools, health centers, and community centers (see figure 6.2).

The involvement of YCI (and its Costa Rican counterpart VIDA) in December 1991 and December 1992 is an interesting example of the influence that such institutions have had on the RSE, and the contradictory assumptions that such organizations make regarding "local communities." Its promotional literature states, "All projects stress local community involvement and ultimately, aim to help people to help themselves" (RSE, n.d.: 31). YCI brings Canadian, Costa Rican, and Australian youths together for periods of several months to work in small teams on defined projects and often in coordination with residents. They raise money to pay for their own materials and supplies, as well as to make donations to the projects on which they work. The initial financial impulse, the design and building of the trails and the interpretation center, and the creation of brochures—indeed most of the international attention the RSE attained—came as the result of the financial and cultural capital of this organization.

Other international volunteer organizations, such as Global Volunteers (GV) that began coming several times a year in 1992 and stopped coming in 2003, have also played key roles in reinforcing the institution.[2] Since then, the RSE's major infrastructural improvement projects have been accomplished mainly through work teams of reserve employees and volunteers, during the two-week-long visits of Global Volunteers. In addition, such groups have brought and sent donations of essential products for the operation of the Reserva, such as books, heaters, concrete for construction, and solar panels. But they are also ambivalently situated within the "local community" they have come to help. They are there to promote "local solutions," based on the assumption that sustainability (environmental, social, economic, and so on) is based upon this.

*6.2. "Youth Challenge International Poster for 'Santa Elena Rain Forest Project,' Circa 1991" (permission of Youth Challenge International).*

And yet the assumption that only an international organization that stays for a short period of time could encourage or secure that is based on a contradictory value that assumes people could or would not do it themselves without the assistance that such organizations provide. Furthermore, in order for these groups to perpetuate themselves, they must ultimately take credit for the work they do, thereby undermining their mission to demonstrate self-reliant communities. In fact, during the decade they were there, GV (mostly unwittingly) insinuated itself into the structure of the RSE. They have sought to erase themselves from this through the depoliticizing language of helping host communities learn self-reliance.

But this is nearly impossible. During 1996, for example, Global Volunteers groups made two different working visits to the RSE. During one of those visits, the team leader told me that GV's mission was to avoid local politics altogether and promote cultural interchange between its clients and locals while helping them value their natural resources. Yet during this visit, GV participants were frustrated that few residents turned out to help them with their projects, which is one of their policies (a local counterpart is expected to work alongside each of their clients). What GV and its participants had not realized, however, was that several months before their visit, the conflict between the Colegio and Fundación had taken place. As one Santa Elenan explained to me, while he and other friends would like to work with the volunteers, none of them wanted to get involved in the RSE right after such a conflictive time. Importantly, the presence of such organizations relies upon and is influenced by ongoing processes of social and political differentiation in the places in which they work; but by assuming unified and homogeneous interests in the communities into which they enter, they tend to reinforce the political and cultural claims and projects of resident leaders and institutions with whom they deal. For example, one leader in Santa Elena who had worked with GV through the RSE sought to influence the organization to send its future groups to institutions and projects in which he was involved. In the past, he had had success channeling GV work groups into non-RSE activities, such as working at a private religious school.

## Making Sense of the Rise (and Demise) of the Fundación

The enhanced status of Juan Montoya and the Fundación in Monte Verde derived in part from their access to institutions like Global Volunteers and Youth Challenge International, strengthening the RSE as an institution with jobs to provide and capital to redistribute. Furthermore, as a link to GV, Montoya was able to direct them to stay in hotels of his choosing, eat at restaurants of his family and friends, and use taxi drivers that he prefers, enhancing his own prestige.

Montoya comes from a prominent landowning family in Monte Verde, and one that stood outside the traditional political structure centered in Town Meeting and the Monteverde Cheese Factory. The RSE represented an opportunity to become involved in the emerging prestige, power, and economic centers of tourism and conservation, not to mention doing something positive for Santa Elenans. Nevertheless, he was deeply ambivalent about the public institution of the Colegio. A month after he quit working at the RSE in 1995, he told me,

> There is a lack of care in the Colegio that has created all kinds of fiscal and personnel problems. How can they improve instruction if they are worried about the Reserva? The other thing is that there is a discontinuity in the Colegio, since teachers and administrators are always coming and going, and they only work nine months of the year. The Junta Administrativa of the Colegio doesn't have the knowledge to run things, and the government is always deficient in how it runs things. Look, the worst farm in the zone belongs to the Colegio. How ironic. They only train *campesinos*. But the thing is there are people here who want their kids to be doctors and lawyers.

As director of the RSE, he was therefore suspicious of the Colegio, and turned to Monteverde 2020, an interinstitutional strategic planning initiative that took place between 1989 and 1994, for administrative advice and support.[3] Out of these discussions emerged the idea to create an organization that would operate the RSE and distribute its profits, within the legal-administrative framework of a *fundación*. Under Costa Rican law, foundations (unlike nonprofit associations) are created to work within or toward a specific mission and can work as for-profit institutions. The mission statement that established the Fundación marked a significant redefinition of the RSE as a community project, promoting that *all* schools (not just public schools) should benefit.[4] To some extent, this position reflects the composition of its board members, only several of whom had children in the Colegio.[5]

By late 1994, Colegio teachers and parents grew restless as they realized their reserve was being run by a new organization that was more closely identified with Monteverde 2020, a private interinstitutional initiative, than with the public institution of the Colegio. Certainly, it is possible that Colegio teachers, administrators, and parents were motivated by the imagined riches of the RSE so that they wanted to bring the project under stricter control of the underfunded school. As one hotel owner joked with me, "It was like an ugly baby at the beginning, and everyone was skeptical that the Reserva would actually make any money. And then all of a sudden, that ugly baby got pretty and everybody wanted to be the father."

And yet, the growing conflict between the Colegio and the Fundación, which ultimately exploded in the protest I witnessed, was not simply an economically motivated move to bring in the riches of tourism. More importantly, it represented the channel for different social and political tensions in Monte Verde that had a lot to do with how the community should be, and was being, defined in the contexts of education, political philosophies around local control, and in styles of political leadership.

One of these issues was clearly the fact that the Fundación was positioning itself not simply as the operator of the RSE, but as an authority through which public discussions on education in the Monte Verde region would take place. This issue, that many Santa Elenans would recognize, debate, and exploit, was of great consequence in the conflict between the Colegio and the Fundación. During the early 1990s, the most active commission of the strategic planning initiative Monteverde 2020 was the "Comité de Enlace—Colegio" (High School Link Committee), whose main purpose was to improve the quality of education in the Colegio. Its most recognized achievement was the design and approval in 1992 of an expansion of the Colegio's technical education curriculum.[6] Its other big achievement was to help establish the Fundación to take over the discussions it had been having over education.

One possible solution to the Colegio's troubles that received significant attention in the 2020 commission, an issue the Fundación inherited, was that of its semiprivatization. To this day it remains a controversial theme, and represents one of the factors that would ultimately unravel the Fundación as an organization. During the early 1990s, the Social Christian government of Rafael Angel Calderón pursued policies of decentralization and liberalization required by World Bank-IMF Structural Adjustment Programs. One public sector that was targeted for a decrease in public investment at the time was education. One high school in the Province of Cartago (San Luis Gonzaga) pioneered a form of semiprivatization, in which a private administration contracted with the state to run the school. Leaders in Monteverde 2020 and the Fundación, inspired by the possibilities this represented for more local control over educational quality, proposed a similar action for the Colegio in Santa Elena. One proponent of the semiprivatization scheme explained,

> The problem of the Colegio's persistent mediocrity was not just limited to Monte Verde, but it's a structural one that is shared by all government institutions. We would contract with the government to run the school, although it would continue to pay salaries. We would have the power to hire and fire teachers, run the school like we wanted to, and respond to the interests and demands of the community. This was all connected to the reform of the state at the time. Of course the teachers at the Colegio opposed this, because they were worried about all the rights they had gained from union struggles. People were asking what happens to the poor who can't pay for private schools, and the like. But there's something basic here, which is when a community has no control over its development, it continues to be disadvantaged.

Even though the semiprivatization scheme was dropped from active consideration by 1993 because of a change in government in San José, and even though members of the Fundación were themselves divided over the scheme, many people who took positions in favor of the Colegio during the conflict did so precisely because of their hostility toward anything associated with privatization.

Furthermore, strategically positioned actors helped shape a public image of the Fundación as a partisan political organization with a neoliberal agenda, even if it did not formally express one itself. A central figure doing this framing was the director of the Colegio during 1995. Doña Cecilia Alvarado was brought to Colegio Santa Elena through the efforts of, ironically enough, some of the Fundación members she would ultimately undermine. Soon after her arrival in early 1995, she argued for a more direct relationship between the Colegio and the RSE, and pressured the Junta Administrativa to take action to bring RSE back to within its own administrative structure. When members of the Fundación insisted on a joint administration of the RSE, she and the Junta began to seek alternative ways to reach the settlement they had been seeking. By June 1995 (when I became involved), they had failed to reach an agreement and so Alvarado began organizing the demonstration. Several months later, Doña Cecilia reflected on her strategies,

> I've had experiences that have prepared me for such a fight, including my hardships as a student in the Soviet Union, and my time in Nicaragua where *machista* men always told me that I was a failure. We had spies all around, telling us what was happening with the Fundación, but when the demonstration failed and they invaded the Reserva the night after, it became all-out war for us. We sent out local and press communiqués. I had some contacts in the national press, and we contacted lawyers and national teachers' union representatives. I went on television because of a contact I had, and here was Monte Verde showing up in a bad light in the press. Nobody liked that, the prestige of the zone was at stake here.

By invoking this struggle as a fight against *machismo*, Doña Cecilia was clearly pointing to underlying tensions over social categories and relationships in Monte Verde, specifically within political processes that tended to admit few women.[7] She admitted, "I'm not like the other women who have power here in Monte Verde. I represent public education, I'm loyal to the youth. They call me communist all the time, even to my face. I'm a revolutionary. I admit it. I won't back down, I'm conflictive where ever I go. I grew up fighting boys and their *machismo*." She was also clearly aware of her position as an outsider who arrived with unusual influence and access in local political matters because of her institutional position.

But Doña Cecilia's concern with fighting the *machismo* she perceived existed alongside her own political philosophy that favored the initiatives of the public sector. She explained to me, "The fear of these Fundación people is that they cannot control public education here in this region. They say the government is mediocre, and sure there is mediocrity in the public education system, but that doesn't mean we have to be." She added, "Local power doesn't work. It's a tradition, not a practice. It's a tradition that's gone, and these people want power for themselves." Her strategies and connections reflected this skepticism of what she called "local power." To help publicize the conflict with the Fundación and to exercise pressure on MIRENEM to intervene, she turned to the lawyers and resources of the Ministry of Public Education and national teachers' unions, seeking to bring the conflict to organizations outside of Monte Verde. But even while she was seeking to centralize authority over the RSE within public institutions, including her own Colegio, she manipulated the language of community empowerment and determination. She observed,

> "Community" is an important theme here in Monte Verde. A lot of people talk about it. But what has that meant here? Three or four people making the decisions. "Community" is consensus, participation. For example, let's say I suggest that we eat, and I want a plate of beans and then you say you want eggs. But you see what's happening here is that your participation is conditional on mine, and it follows and speaks in terms of mine. With the Reserva, three or four made the decisions and the rest followed their orders. That's not "community." But this is how things work here, they work by *amistades* [friendships] and connections, unless you have popular power.

During the demonstration in which I participated, she encouraged the assembled crowd to unite as a community to confront the crisis with the Fundación. But she was not assuming an already united or harmonious locality. She knowingly channeled social and class tensions by retooling the language of local power and community initiative into a format that sought to confront head-on the powerful groups that controlled those processes locally. In her strategic position as director of the Colegio and with her substantial charisma she was ultimately able to encompass the language of community control within a wider mission of strengthening the Colegio, and through that, centralized authority of the Ministry of Public Education in Santa Elena.

Another strategically positioned actor involved in the debate was the Catholic priest of Santa Elena, who took an active role in mediating the conflict between the Fundación and Colegio. His views represented an influential position for many residents of Santa Elena who took a position against the Fundación during the conflict, even though he took no public stand on the issue. As he told me, "My role was to call the two sides to dialogue, to sit at the same table and work this out without defamation or threats. Defamation and threats were coming from all sides. So this was the official position of the Church, to have the two sides work this out in civility." He continued that his personal perspective was quite different, though, and he was quite open about the inequalities of power and access to resources that the Fundación represented. He challenged the political and social authority of the Fundación:

I don't know if you have heard a term we use here, "*cacique*." It's the system that has existed here a long time in Costa Rica. Perhaps it is less so now because people are waking up to their own power. But I think it is possible to see in these people of the Fundación possible *caciques,* the same people who have directed the town for years. This struggle for the Reserva was urgent, because it was said that these people in the Fundación wanted to privatize the Colegio. I ask you: what does this mean to the poor *campesino* who cannot pay to send his child to school? They say there are grants for these kids, but how does a kid who only eats rice and beans, maybe not even breakfast, compete with the children of these people? You can see these kids with their long faces, they fall asleep in classes. It will perpetuate the same political system if we educate only a few and leave the rest uneducated. This was a matter of social justice.

The priest emphasized that the Fundación's control over the RSE perpetuated social injustices and class inequalities.

My neighbor Don Roberto was one individual who seemed influenced by this perspective, describing the Fundación as an organization to which he and others who were not themselves already integrated into Monte Verde's elite political processes were only partially invited. He once explained to me his participation in the negotiations and demonstration, as a member of the Junta Administrativa:

These people want to privatize the Colegio. Now what does that mean to someone like me, who is poor? It means I can't send my daughter to school. I can't afford a private education here, neither can most people. This Fundación was formed without the Colegio. But we have been powerless in the face of the slow and unmotivated bureaucracy of the government. We want the Reserva to pay for good teachers, and support students with materials. The government has been unresponsive, even against us, this whole time as we sought to retrieve the Reserva. What the people need, Luis, is the power to decide and control their own destinies.

What is most striking about this perspective is that, like the arguments of Fundación members, it asserts the need for some form of local control and decision-making process over the Colegio's administrations. And yet these people ended up on opposite sides of the conflict over the RSE. The difference is that for Don Roberto and other Colegio partisans, in spite of their own references to their self-determination, recourse to the Costa Rican state was a legitimate strategy to solve their conflict.

One way to explain this contradictory tactic is that Don Roberto was challenging one style of political leadership represented by the Fundación with a self-consciously populist style of leadership. The processes that undermined the Fundación were based on the emergence of individuals and groups who did not participate in consistent ways in the normal political processes of Monte Verde. In Santa Elena there is a term for these individuals: *caciques de pleito* ("fighting chiefs"). This style of leadership is characterized by seemingly spontaneous popular coalescence around issues and organizations, encouraged and facilitated by key individuals, that follows the lines of existing (and creates new) social and political alliances, if only briefly. For example, the Asociación de Padres de Familia of the Colegio (the Parents Association, like the PTA), mainly dairy farmers, agriculturists, and small business owners, emerged as important actors in siding with the Colegio Junta Administrativa and administration. Its members brought with them alliances through family, marriage, and friendship, and were able to bring significant moral pressure upon Fundación members to renounce their claims on the RSE. Several of its members also played key roles in organizing the demonstration at the RSE. Soon after the crisis, however, many of these people returned to no or less obvious forms of activism.

In some cases, the rhetoric of these *caciques de pleito* referred to Fundación members as *"gente de afuera,"* or outsiders, though by this they did not necessarily mean foreigners or non-Costa Ricans. They represented the Fundación as the convergence of newly settled residents from other parts of the country (as well as one North American man living in Monteverde village), noting that behind Fundación members' rhetoric of neoliberal decentralization lay the erosion of their own political claims, access to public resources, even cultural values. One farmer observed, "These people who ran the Fundación were all from the outside. This is why there were problems, because all they cared about was making money. They're the same people who have prevented the road from getting fixed properly. They're all in tourism. Here we are, grandchildren of this town, and we live poorly, and no one wants to work the land, they all want to work in tourism." Though this assertion is neither necessarily accurate (some Fundación members were born and raised in Monte Verde and not all worked in tourism), nor is it shared by all critics of the Fundación, it reflects both individual and collective concerns over the rapid social and economic changes in the region as a result of tourism and environmental conservation – forces that themselves are seen (at least by this person) to come from the outside. In this case, definitions of outsider reflect relative positionings vis-á-vis the person who is making the claim.

In fact, a number of people on both sides of the conflict suggested to me that at the root of the conflict were changes in the structures and styles of communication between residents as a result of the demographic changes and pressures brought on by the tourism economy. One observer saw it in this way: "Perhaps something that is important here is that the arrival of new people looking for work has changed the way people communicate here. Before, people would run into each other on the roads, at the *pulpería,* and there they would exchange ideas and thoughts. But the new people that have come don't automatically fit into this communication system. Also, nowadays everyone is working so hard that no one has time anymore for visiting and socializing. Economic concerns have imposed themselves on people's lives now." He added that the recent economic prosperity has allowed people to buy motorcycles, which are an antisocial means of transportation, since they allow people to avoid social contact on the road.

## Conclusion: The "Entanglements of Local Politics"

On 1 September 1995, the Colegio took full administrative control over the RSE, and it has maintained firm control since then. Several weeks after the demonstration at the RSE and the media attention it generated, the Vice Minister of MIRENEM called a meeting of the main protagonists in Santa Elena. The stated policy of MIRENEM during the 1994–98 government was that environmental conflicts should be settled within community and municipal contexts (Castro 1996), a policy that has encompassed the rhetoric of local control and administrative decentralization while recognizing the budgetary realities and constrictions of the Costa Rican state. Nevertheless, it was prodded to intervene since as the owner of the land, it had the final say on who should administer its land. Since the Colegio's Junta Administrativa had the legal concession to the RSE all along, something the Fundación had never secured, MIRENEM required that the Fundación forfeit its administrative role. Since its specific purpose was to operate the RSE, the Fundación dissolved.

Two particular perspectives on this conflict reflect on an issue of wider relevance. One is the observation of the MIRENEM official assigned to oversee the RSE after the Colegio formally regained it. When I interviewed her, she explained that she never understood why the conflict happened, when it came about, or even who the major actors were: "It's the entanglements of local politics to me." The other was the reflection of my neighbor Don Roberto, who said, "You know, none of us ever understood how we reached a settlement with this conflict. It was out of our hands the whole time since MIRENEM was the one who really held the cards and was working behind the scenes, and then suddenly it was solved by the Vice Minister." In the former observation, the local community is viewed as a distant abstraction characterized by its opaqueness, the force that determined how events unfolded. In the latter, poorly understood dynamics beyond Monte Verde are seen to determine the form the conflict took and the settlement reached. They each represent selective and partial understandings of the conflict over the RSE, but they share in common the assumption that what was really going on in the RSE conflict was beyond their vision and comprehension, even if they were involved at various stages of the conflict.

It is in these spaces of situated comprehensions, desires, and their disjunctive relationships that the meanings of "locality" and "community" are debated, negotiated, imposed, and realized. It is also here that locality takes on a fetish quality (Appadurai 1996), obscuring its ultimate referent, becoming something seemingly tangible from a distance, and yet at the same time, inaccessible. This is seen most clearly in the comment of the MIRENEM official, for whom locality represents an identifiable entity, even if it is entangled; and yet its precise elements defy identification. It is also here that YCI and GV unreflexively work for community self-reliance. But this fetish does not have the same effect for everyone who engages in the discourses and practices of locality. An easily identifiable local like Don Roberto (whose family is among the first settlers of Santa Elena) may attribute true causes and resolutions for the crisis in which he was involved elsewhere, outside of his own immediate realm of social action. So he has no problem selling the label of locality—RSE as a community project—to an organization like Global Volunteers since it brings in economic resources for what he considers an important cause, and allows him to communicate to the people with whom he lives the desire he has to share with them in common cause. For him and others, conflicts over the RSE have not simply represented the frictions over who manages a natural area best, class conflict, or tensions of different styles of political leadership in Monte Verde, but debates over the concrete futures of their children, and who should be in charge of their educational processes.

In defining and redefining the parameters of their community project, Montoya, Don Roberto and others, along with Doña Cecilia, have turned the category and attributes of community on its head again and again, by seeking beyond the shifting boundaries of their immediate geographical and social communities to resolve their crises, if not to also develop their cloud forest reserve. Turning again to the theme of "local communities" within conservation planning discourse and practices, it is not possible to assure that the theme of locality is integrated into any conservation initiatives in constant and universal ways in places like Santa Elena. Seeking to identify the multiple forces and organizations that transcend and constitute the locality in question only treats a single aspect of the problem, at the risk of essentializing those forces and frameworks. This uncertainty lies in the multiple meanings of the very concept of local community, a concept vague and flexible enough that it can be appealing and useful to a variety of interests in specific circumstances.

## Toward a perpetual (post)script

About six months after the Colegio regained administrative control over the RSE, there was a meeting in which members of the Junta Administrativa of the Colegio, teachers, and RSE employees began working on a five-year renewal of the concession on the RSE land. Significantly, Doña Cecilia was not present, and she would not return to the Colegio as its director. Many people took credit for this fact, including members of the same Junta Administrativa of the Colegio that had supported her throughout the conflict with the Fundación. Within a month of regaining the RSE, they entered into their own struggles with her personal style. Others suspected that the politically connected members of the Fundación and their allies had gained their revenge on her and arranged at top political levels that she not return.

As we reviewed the 1990 proposal that originally outlined the ecotouristic-investigation project, the new administrator of the RSE, the second in the last six months (and one of many who would follow), asked the question that rested just barely under the surface of any discussion over the RSE: "I'm curious about this being called a community project. I think we need to examine that. Who is the community here? How do we define it?" One of the members of the Junta Administrativa muttered in an exasperated tone, "*Otra vez.*" ("Again," as in "here we go again!")

# Notes

1. A non-exhaustive list of supporters who offered financial, material, and professional resources between 1991 and 1993 includes Monte Verde institutions (the Cooperativa Santa Elena, a supermarket, the Monteverde Cheese Factory, Cámara de Turismo de Monteverde, the Instituto Monteverde, hotels and others for material contributions and for arranging to send international volunteers to work there), national and provincial government agencies (the Municipality of Puntarenas, who built the road from Santa Elena to the RSE, and the Instituto Costarricense de Turismo, which provided two million *colones* to pay for it), and international organizations and individuals (such as independent volunteers, Youth Challenge International, and Global Volunteers). Furthermore, in 1985 there was a technical review process sponsored by the Ministry of Agriculture in which the Finca Río Negro was redesignated for purposes of forestry, not agriculture (MAG 1985).

2. Global Volunteers calls its programs "service learning travel" in which (mainly) Northern individuals volunteer in towns and cities in the U.S. and other countries in programs ranging from infrastructural projects and professional services to teaching English, in order "to establish working partnerships with communities striving to become self-reliant" (Global Volunteers, 1996). Generally, though not necessarily, work groups consist of ten or so older North Americans with a team leader, who stay in "culturally appropriate" hotels or private residences. Depending on the site, trips range from $350 to $2,000 for one or two weeks. A small portion of that money, between $25 and $150 per participant, is donated directly to the "community projects" in which they work (ibid.)

3. The strategic planning initiative called Monteverde 2020 (formed in 1989 and lasting until 1994) brought together the representatives of different regional institutions to discuss the issues surrounding the question "What should Monte Verde be like in the year 2020?" and to develop strategic plans and options to accomplish goals related to improvements in economic sustainability, conservation, and education. It received funding from the Interamerican Foundation and institutional participants.

4. The mission statement says, "The Fundación has as its ends the improvement, using the funds its activities generate, of the education and quality of life of the community of Santa Elena de Puntarenas and neighboring places. The improvement of education will be realized through the support and benefits that the Fundación provides to the schools and high schools of Santa Elena and neighboring zones; the training in languages, tourism, administration, biology, and related matters that this offers the people of the community. The improvement of the quality of life will be realized with the programming and defining of strategies, through which people of the community will offer the services of ecotourism, transport, guides, rental of horses, hotels, alimentation, and other touristic services; so that they benefit economically from the activities of tourism in the zone" (Fundación 1992).

5. As required by the statutes, the board of directors of the Fundación included representatives of the Asociación de Desarrollo Integral de Santa Elena (Santa Elena Integral Development Association, established in 1977, works as the liaison between municipal and national governmental authorities and Monte Verdeans, organizing infrastructural and development projects, mainly road maintenance), the Asociación de Padres de Familia of the Colegio (The Parents' Association of the Colegio, which ensures that basic services, such as the cafeteria, continue to function in the Colegio), the Junta Administrativa of the Colegio, the Municipality of Puntarenas, and VIDA (the Costa Rican counterpart of YCI). Furthermore, several managers at the Monteverde Cheese Factory agreed to be board members, bringing a level of political prestige, since Cheese Factory managers have often been in positions of political influence given the economic importance of their institution.

6. Monteverde 2020 successfully developed a plan to add a tourism component to the curriculum, so that students could learn English and skills to work as hotel receptionists and administrators, nature guides, and travel agency workers.
7. This is not to say that women have not participated in political processes. The former governor of the province of Puntarenas (in which Monte Verde is situated) is a woman who lives in Santa Elena, and has been involved in political processes in the region for many years.

# III Part Three: Monte Verde and the Adolescence of Ecotourism

During 2004, while I was a Fulbright scholar at the University of Costa Rica, I was invited to participate in Monte Verde's first annual "Expoferia Ecoturística de Monteverde," or Ecotourism Expofair. Sponsored by the ICT (Costa Rican Tourism Institute) and CETAM (the Monteverde Tourism Chamber), the three-day fair "Naturalmente Nuestra" ("Naturally Ours") was an opportunity for the region to showcase its tourism offerings to journalists, industry professionals, and tourists. To kick off the fair, the local planning committee (a collective of tourism entrepreneurs, conservation groups, and other area institutions) had organized a panel to take place on the fair's first day, called "Ecotourism and Sustainability in Monteverde."

The Monte Verde tourism sector has long described itself as existing "in harmony with the cloud forest," this being a phrase that a number of businesses there use. Scholars have also regularly declared ecotourism there as a beacon of sustainable practices in the global tourism industry. As a pair of Costa Rican economists recently argued, "The experience of Monteverde, one of the most important ecotouristic destinations in the country, demonstrates that in Costa Rica there is a great potential for communal environmental responsibility. The community of that place should be an example of ecotouristic development on the basis of which governments should promote sustainable competitive strategies for the national tourism sector" (Acuña and Villalobos 2001: 10, my translation).

I was a little nervous about presenting my own views, which are more critical of the image of a tourism industry living in harmony with the cloud forest. But when the first panelist, a high-ranking administrator at the Monteverde Cloud Forest Preserve, expressed at the outset of our panel that the local version of ecotourism is neither ecologically or socially sustainable, I understood that this panel was not set up as a self-congratulatory exercise, but as an internal critical analysis of what tourism has wrought. Indeed, as he and the other panelists

presented their (sometimes harsh) criticisms about the lack of control in the
tourism sector's growth, it became clear that there was widespread
preoccupation that the tourism industry, an important justification and funding
source for nature conservation (Honey 1999), had itself become a threat to
Monte Verde's distinctive ecology and social dynamics. Instead of being one of
the most positive encounters with environmentalism—as expressed by the
economists above—it was becoming one of the most threatening.

It was not just the scale of tourism that worried panelists (local tourism
leaders were projecting two hundred and thirty thousand visitors in 2004, up
from sixty thousand just a decade before), but the fact that hotels, restaurants,
tourism attractions, and worker housing had been built so rapidly and without
regulation that erosion, water contamination, deforestation outside the reserves,
and urban crowding were reaching unacceptable levels. Some panelists worried
that Monte Verde's very identity as an iconic place of nature was at stake, while
others expressed concern over growing socioeconomic inequalities generated by
tourism. For example, land prices in the Santa Elena-Monteverde tourism
corridor have shot up (as much as $15–20 per square meter, comparable with
costs in the capital of San José; Chamberlain 2000: 376), making it difficult for
working families to find acceptable housing close to their places of work. They
said that tourism had deepened class divisions and undermined traditional styles
of conviviality as residents competed with one another and treated their visitors
as inhuman sources of dollars. If things go poorly in the next few years, several
panelists warned, even the formally protected areas could come under threat
from squatting and social unrest.

The problem is more acute when one realizes the extent to which Monte
Verde's economy has come to rely on tourism. Economists observe the presence
of an "ecotouristic cluster" there, that is, a concentration of businesses in which
the competitive success of each one depends on the others (Acuña and
Villalobos ibid.). There are currently over fifty hotels and *pensiones,* with an
average of eighteen rooms per establishment, up from thirty hotels just seven
years ago, and a growing number of businesses (hardware stores, groceries, etc.)
that serve the tourism sector. Tourism has clearly replaced dairy farming in the
Santa Elena-Monteverde corridor, with only a handful of dairy farmers left.
Although accurate numbers are difficult to come by (economic statistics are
scarce in Monte Verde), a recent study approximates that during a typical visit
(two nights, three days) in 2002, the average visitor spent $63 per day on such
services as hotel, restaurant, horse rentals, transportation, and souvenirs
(Sanchez 2002: 27). Although it is likely that this number is low (the ICT
reports average tourist expenditures of $85.50 per day across Costa Rica), it
generates a conservative estimate of an economy directly related to tourism of
some $14 million (based on projections of two hundred and thirty thousand

tourists a year). The number of people directly employed in tourism is about eight hundred, and some four thousand are thought to be employed in businesses that rely indirectly on tourism (ibid.).

Monte Verde's tourism sector is characterized by local ownership of small-scale businesses, unlike other parts of Costa Rica, where ownership may be foreign or corporate (Pérez 1999; Ramirez 2004). As a result, it has had a comparative advantage in that tourism revenues are not as likely to "leak" as severely as in other tourism destinations.[1] The rough condition of the road—still not paved to the Panamerican Highway thirty-five kilometers away, even as Monte Verde occupies a top spot in the Costa Rican tourism industry—is one reason large-scale investors have stayed away, because it makes the costs of doing business there higher.[2] The tourism industry that has emerged in Monte Verde, then, can be rightly called "home-grown." It is largely operated by people who grew up in Monte Verde or nearby rural hamlets, and a respectable number of them are now wealthy and politically connected hotel or restaurant owners. As an economic sector, they have declared their abiding respect for nature and its conservation, and regularly donate time and money to support community initiatives, such as the fair I attended. There is no doubt that they have benefitted from and helped shape Monte Verde's international reputation as a premier nature destination.

For these reasons—that tourism is based on small-scale and local ownership, led to the empowerment of locals, generated local respect for nature, and so on—Monte Verde is often showcased as more than a merely "nature-oriented" destination: it is an *ecotourism* destination, indicating something more profound (Honey 1999). According to The International Ecotourism Society, a major global promoter of the concept, ecotourism is "responsible travel to natural areas that conserves the environment and sustains the well-being of local people" (Wood 2000: 1). Ideally, it is tourism that minimizes impacts, provides direct financial benefits for local communities, empowers local people, raises sensitivity about host countries, supports human rights, and builds environmental awareness among both hosts and guests (Honey 1999; TIES 2004). One of its most important effects on national development is that it helps generate economic activity in once peripheral rural areas (Fürst and Hein 2002). The Monteverde Cloud Forest Preserve has been widely promoted as an example of how ecotourism can directly support nature protection and serve "sustainable development" (Tobias 1988; Tobias and Mendelsohn 1991; Chamberlain 1993; Aylward et al. 1996; Acuña and Villalobos ibid.). While foreigners account for upwards of 80–90 percent of MCFP visitors (some fifty-five to fifty-seven thousand visitors per year), they have contributed almost all of the revenue raised from entrance fees, providing an 8 percent return on investment (Aylward et al. ibid.: 328–9; Figueroa 2002; Sanchez 2002).[3] These profits ($420,000 per year) have allowed the MCFP to invest in environmental education in local schools, help out with road maintenance, develop a recycling program, and sponsor cultural and ecological fairs.

Twenty years ago, visitors to Monte Verde had to work relatively hard to get there, and once there, the main, if not sole, attraction was the area's protected cloud forest. Visitors were typically committed bird enthusiasts or amateur naturalists, their main practice being quiet observation of flora and fauna, and their accommodations were generally simple. Today, there are a number of high-end luxury hotels, and about half (52 percent) of Monte Verde's visitors come on package tours on private air-conditioned buses, arranged through travel agencies (Sanchez 2002). The number of area attractions has blossomed to include adventuresome "canopy tours" (bridges and zip-lines through the trees), a Tarzan swing, paintball gun battles ("Let's play in the forest!"), golf cart rentals, and rent-a-quadracycle, among others. Although their promotional materials all emphasize that these activities are good ways to know the forests and biodiversity, if they are not also good for conservation itself, some residents have become uncomfortable with the label "ecotourism" to characterize what is going on there, or "ecotourist" to describe the distinctive people and sensibilities attracted to these activities. Part of the problem—one that promoters like The International Ecotourism Society have become intensely aware of as all manner of mischief takes place under the label they promote—is that greenwashing (claiming a rhetorical commitment to ecological and social sustainability while in practice undermining these things) is common and that there is little accountability within the industry (Honey 2000). It is such concern that is driving a current global movement to establish standards and certify tourism businesses as "sustainable" (the Costa Rican government has its own Certification for Sustainable Tourism), although such issues have yet to penetrate much into Monte Verde.

This is the context in which panelists expressed their preoccupation with ecotourism's relationship to sustainability, and declared the tourism industry's ambiguous consequences for Monte Verde. Notwithstanding the fact that "sustainability" is a twisted and vague concept that nobody defined that day (Mora 1998), what their concerns most clearly demonstrate is that narrow perspectives on ecotourism that define it solely in terms of its economic productivity for conservation and sustainable development miss the bigger questions a number of Monte Verdeans themselves are asking about the quality of their lives, the social impacts of tourism, the ecological consequences of hosting several hundred thousand people every year in a fragile territory, and the unevenness of its benefits and riches. As this and other examples of the abuses or unfulfilled promises in the name of ecotourism indicate, it seems clear that ecotourism has moved out of its childhood, a period marked by triumphant and prescriptive discourse, and entered a more complex, even uncertain adolescence (Vivanco 2002c). In the euphoria of its emergence and rapid niche market expansion as a conservation and development scheme, certain patterns, some of them contradictory, have persisted and been amplified in places like Monte Verde, raising important questions as to ecotourism's much-touted status as an "alternative" form of tourism (Cater 1994). It is likely that as ecotourism's adolescent years bring new crises of self-definition and maturation, its foundational contradictions will intensify, albeit in distinct ways in different contexts. To fully appreciate the implications of this situation requires a reorientation from a polarized discourse of, on the one hand, prescribed acceptance of ecotourism by its policy-oriented proponents (i.e., Boo 1990; Lindberg and Hawkins 1993; Honey 1999; TIES 2004), and on the other, outright rejection by its critics (i.e., Kamuaro 1996; McLaren 2003; Vivanco 2002c).

This section of the book contributes an ethnographically grounded analysis of ecotourism discourse and practice (Bandy 1996; Vivanco 1999; Belsky 1999; West and Carrier 2004), by focusing on how the Monte Verde tourism industry has helped shape how visitors and residents alike think about and interact with tropical nature. Often lost in the global policy-oriented debates over ecotourism are fundamental sociocultural questions about how it actually operates on the ground, where its presumed distinctiveness often melds into more traditional forms of "mass" tourism. My contention in the chapter that follows is that in its objectifications of people and nature, ecotourism canonizes certain versions of natural and social history, to the advantage of some, the expense or silence of others, and the considerable reduction of complexity and heterogeneity (Horne 1992; Hollinshead 1999). Encountering Monte Verde landscapes and communities as a tourist offers a performative means by which certain themes about how to manage unpeopled tropical landscapes can be explored. But that performance exists in an often unrecognized tension with the human communities that live in the rural landscapes that are the object of ecotouristic desire and attention. The result is an othering of both nature and community, as nature and human relationships to it are increasingly being reconceptualized and normalized through the market-oriented domain of ecotourism. In many ways, it is appropriate to begin concluding this book with this theme, because when many people—environmental activists, *campesinos,* and business owners alike—describe Monte Verde's (not to mention Costa Rica's, see Fürst and Hein ibid.: 510) future in the era of reduced international funding for environmentalism, ecotourism's vision and practices are given central place in spite of increasing concern over its ambiguous effects.

# Notes

1. On average, 70–80 percent of every tourist dollar spent on a trip flows not to communities but to cosmopolitan airlines, travel agencies, hotels, etc. (Third World Network et al. 2001).

2. One of the access roads from the Panamerican Highway (there are two) is in the process of finally being paved, although recent financial scandals involving the company that is paving it interrupted it in 2004. Hoteliers have been a key force in promoting the paving of the road, although common knowledge around Monte Verde is that they stand to lose a lot from a paved road since tourists would no longer stay several days, as they do now (the average visitor stay is two nights), but come and go within a day.

3. As one MCFP document states, "International tourists, if properly motivated, represent a major source of revenue both during their visit and afterward. Follow-up funding from tourists subsequent to their visits has the potential of becoming a major source of support for Third World conservation efforts" (Aspinall et al. 1991: 7). The MCFP has sought tax-deductible donations from visitors through the U.S.-based nonprofit support group called "Friends of the Monteverde Cloud Forest." It has not provided substantial resources, contributing only 4 percent of MCFP revenues in 1993, for example, (Aylward et al. ibid.: 329).

# 7 Quetzals and Other(ing) Spectacles of Tropical Nature

On the morning back in 1995 when partisans of the Colegio marched on the Reserva Santa Elena, marking a high point in public outrage against the Fundación, it was, to say the least, an interesting day to visit the small cloud forest preserve. It was a colorful and raucous protest march, and people were speaking earnestly out in the open about important issues that bear on nature conservation, tourism, and public responsibility. Even though it was not tourist high season at the time, the RSE could have expected at least twenty visitors that day. But word had gotten out, at least to the naturalist guides and hotel owners, that the protest was going to happen, and only two North American tourists dared show up. Even more striking was the almost total lack of curiosity they showed about what was taking place before them. The naturalist guide who accompanied them, known to virtually everyone at the protest, was eager to usher them into the forest, pausing only to point out a specific plant near the reserve's entrance, barely acknowledging all of us gathered there.

I do not know if these people thought of themselves as "ecotourists," "nature tourists," "sustainability tourists," or for that matter, just plain old "tourists." These labels (minus the non-modified "tourist") have been widely promoted as low-impact forms of adventure travel in support of natural resource conservation (Boo ibid.; Fennel and Eagles 1990; Whelan 1991; Bandy 1996; Honey 1999). Ecotourists in particular tend to see themselves as key actors in generating sustainable economic development for local communities, new worldwide appreciation for nature conservation, and social equity. But there is ambiguity in such labels. While ecotourism professionals and their allies in the development industry work hard to distinguish the concept from other closely related forms of tourism (especially "nature tourism," see Honey 1999), in practice travelers, entrepreneurs, and government officials rarely make ecumenical distinctions, and the label "ecotourism" serves as a catch-all word that businesses and visitors alike often use to describe what they are up to in Monte Verde.

The lack of tourist interest or involvement in the protest that day was a missed opportunity for visitors to encounter the persistent real-life dilemmas of nature conservation, and symbolizes ecotourism's ambivalence about the actual communities in which it is practiced. The fact is, the profound conflicts and social complexities that were the reason for the protest are rarely acknowledged in the grounded practices of Monte Verde ecotourism. Social communities and their histories are often rhetorically bounded, homogenized, and stereotyped, related to the perception that "local politics" like those happening at the entrance to the RSE are marginal, if not irrelevant, to ecotourism's priority of seeking out and experiencing tropical nature. This is based on a number of unexamined assumptions, including, for example, that while ecotourism emphasizes learning and self-improvement (Munt 1994), such conflicts can be a downer on that other crucial aspect of ecotourism, the pursuit of pleasure and spectacle. Or that wilderness is by essence unpeopled, so that locals, especially rural Costa Ricans, can represent an awkward interference to an unmediated nature experience. They are often identified in tourism (and environmentalist) narratives as destroyers of nature itself, although they are increasingly constructed as key partners in the effort to save tropical nature. When people are relevant, because ecotourists *do* train their attention on human communities, some people (such as Monte Verde's famed North American Quakers and scientists) are more appropriate to see than others.

The force of this ambivalence does not rest on essentialized images of destructive Third World peoples and peaceful nature-respecting Quakers alone, however, for ecotourism is also an arena in which powerful ideas and images of tropical nature circulate. Indeed, for many Monte Verde visitors, the very power of viewing and experiencing tropical nature in a cloud or rain forest derives from its perceived disconnection from the rest of humanity, a central goal being to engage with the spiritual transcendence, moral neutrality, or utter incommensurability of nature itself (see also Davis 1996). Some analysts have argued that this engagement with nature's otherness is one of the factors that self-consciously distinguishes eco- and nature travelers from other forms and categories of traveler and tourist (Wilson 1992; Bandy 1996). But as Craik (1997: 113) has observed for the category of "culture" in tourism, "nature" also has multiple phenomenological and sociological valences, simultaneously existing as resource, product, experience, and outcome. Furthermore, in its specificity tropical nature also represents habitat, refuge of inspiration, playground, wilderness, Eden, jungle, and laboratory (Wilson 1992: 12). For the consumers and producers of Monte Verde tropical nature, tourism therefore represents a negotiation over not only the otherness of tropical nature but how to organize and make sense of the encounter with that otherness.

Importantly, touristic encounters with tropical nature tend to exist in a cultural context that Kirshenblatt-Gimblett (1998) calls "fragmentation," which is to say that the very logic of exhibition is to detach things from their contexts and to reduce lived-complexity. Tourism objects and experiences are strategically juxtaposed to create and reinforce certain story lines, reflecting the themes the producers of the experience want to advance. These mediations are vital to the production of a tourist attraction as both a space of posited meaning and a space of abstraction, because it is from these juxtapositions that new meanings are derived and relationships that otherwise would not be seen are revealed (ibid.: 3; Horne 1992). In Monte Verde ecotourism, this has meant an emphasis on certain foundational themes—tropical nature is inherently separate from human domains, it is threatened, it is a space of mystery whose secrets can be unlocked through science, and the very act of touring the forests is helping save them. These themes are crucial for helping ecotourists understand how their activities fit into a wider global narrative of nature's redemption, but as Kirshenblatt-Gimblett herself (1998: 21) rightly points out, "There are as many contexts for an object as there are interpretive strategies," and it should be added, levels of self-reflexive engagement in the negotiation of authorized story lines.

Within the context of ecotourism, Monte Verde is an idea, even an ideal, but it is also an actual place whose touristic spaces are organized in ways that reflect, shape, and diverge from the idea(l)s. Tourists come to Monte Verde in search of certain things and images—cloud forests, resplendent quetzals, genuine grassroots conservation, and so on—acting out, confirming, and sometimes challenging story lines that focus on the region's conservation and development successes. This has been highly effective for the region's economic productivity, so even while it has displaced other economic activities (dairy farming), it has also created new ones in the process, attracting so many new migrants to the region that the population of Santa Elena is six times what it was just twenty years ago. But also at the heart of these processes exist contradictory messages and situations, one of those being, for example, that even while ecotourism seeks to bring nature and communities closer to the tourists, tourists and tourism producers engage in certain kinds of objectifications that confirm the otherness of these categories. There are also basic tensions, for example, between ecotourism's tendencies to turn tropical nature into a visual spectacle and space of adventure, and the scientistic efforts to place rationalistic order on it. These issues are becoming more apparent as ecotourism, once a relatively small-scale endeavor in Monte Verde, spawns a whole new set of offerings and sensibilities—canopy tours and other adventuresome activities—whose connections to issues of biodiversity conservation and sustainability seem more tenuous. Ecotourism has become an increasingly crucial site of power and contestation within Monte Verde environmentalism, and not just because ecotourism represents a key domain wherein environmentalist narratives and concerns can influence the shape of economic development in the region. It is also because environmental organizations like the Monteverde Cloud Forest Preserve and the Santa Elena Reserve have used it as a key strategy for their own financial survival, and so the opportunities and dilemmas of ecotourism are in turn helping to shape institutional priorities and decisions.

## Monte Verde and the Rise of Ecotourism

Monte Verde's status as a prestigious ecotourism destination is inseparable from Costa Rica's stature as a key site of "scientific tourism" (Laarman and Perdue 1989). Since the founding of the Organization for Tropical Studies in 1963, scientists-in-training as well as senior researchers have been coming to Costa Rica (and of course Monte Verde) to participate in tropical biology courses and to undertake research. Laarman and Perdue, in their analysis of Costa Rican science tourism, assert that the country's exceptional biological diversity within a small area, comparatively strong system of parks and protected areas, relative proximity to the U.S., and green image in international media have made it especially attractive for scientific research (ibid.; Green and Barborak 1987). Scientists, they argue, were the first "trail blazers" of Costa Rica's nature-based tourism industry, promoting the country by word of mouth as a pristine destination.

Some analysts have also pointed to Costa Rica's reputation for not having an army and Oscar Arias's receipt of the Nobel Peace Prize as important stimulants for tourism during the 1980s, because they established Costa Rica as a safe place for North American investors and tourists in a conflict-ridden region (Rolbien 1989; Honey 1994a; Honey 1999). This media-fed reputation in North America coincided with a profound economic transformation in Costa Rica, in the context of structural adjustment programs. As the government reduced its support for previous modes of economic development in peripheral areas (largely agroindustrial enterprises), it searched for new sources of income and investment, and began promoting tourism through infrastructure development, marketing, and regulatory incentives, such as the 1985 Law for Tourism Incentives (Fürst and Hein 2002). Although tourism surpassed coffee and bananas in 1995 to become the primary earner of foreign exchange (employing directly or indirectly some 15 percent of the economically active population), the expansion of the nontraditional agro-export sector and *maquiladoras* (such as Intel computer chip assembly) have offset any national dependency on tourism (ibid.). Since 1999, Costa Rica has been undergoing a boom in foreign visitation (even with problems related to 9/11; see Báez 2001), as new large-scale complexes along the Pacific coast have come into existence. Recent tourism development has generated severe problems of water contamination, waste management, lowered water tables, and deforestation, especially in beach areas (Fürst and Hein ibid.: 129), and in 2004 the Minister of Environment and Energy declared the national park system to be in a "state of emergency," primarily because of 1) ongoing state fiscal crisis, but more importantly 2) tourist businesses that have profited from the natural areas have not been investing back into their protection, especially in national parks (Barry 2004: 1; García 2004). Yet Costa Rica's identity as a green destination is stronger than ever, and some 45 percent of tourists report flora and fauna as key reasons for their travel (Borges 2004: S–2).

As a prominent and early destination of scientific tourism in the 1970s and site of Costa Rica's first private protected cloud forest, Monte Verde became known as an excellent place to view certain iconic species and to interact with cloud forest ecosystems. In the early years accommodations were mainly in the Quaker village, and it gained a reputation as a hospitable community among the mainly North American visitors. Visitor numbers grew gradually throughout the 1970s, and the 1979 *National Geographic* film "Forest in the Clouds" increased the area's international profile, so that by the mid 1980s, the MCFP (the only real attraction during that period) recorded sixty-five hundred visitors. Monte Verde's first real boom in tourism, in which attractions diversified and people outside the Quaker community began to become more directly involved in the activity as hoteliers, restauranteurs, taxi drivers, and naturalist guides, took place in the late 1980s and early 1990s, as Costa Rica's own tourism numbers rose, and Monte Verde was becoming a key icon within transnational "save the rainforest" efforts. Positive reports about children saving the rain forests helped feed a 41 percent annual increase in tourist numbers between 1989 and 1992 (Aylward et al. 1996). As Costa Rica's tourism industry has undergone another boom in the 1990s, so has Monte Verde's. The type of visitor to Monte Verde has also diversified in the past decade, as the percentage of tourists who come on tours set up by travel agencies rose to over 50 percent (Pérez 1999: 51; Sanchez 2002).

Throughout this history, it is noteworthy how media and scholarly representations of Monte Verde have helped shape it as a key ecotourism destination in a global pantheon that includes Kenya, Nepal, Belize, and others. So another important aspect of Monte Verde's rise is the fact that ecotourism scholars and advocates have long pointed to it as a successful example of and justification for the concept (Wearing 1992; Aylward et al. 1996; Báez 1996; Báez and Valverde 1999; Honey 1999; Wearing and Neil 1999). This has itself generated a certain amount of scholarly attention, attracting researchers (like myself) interested in studying its operations (cf. Grosby 2000: 376) and promoting its effectiveness (Wearing ibid.; Aylward et al. ibid.; Báez and Valverde 1999). It is difficult to estimate how much tourism is generated by such scholarship, although it is likely that the scholar's contributions to global media discussions of ecotourism and their influence in transnational ecotourism NGOs (such as the International Ecotourism Society) have helped strengthen the perception that Monte Verde has succeeded in its quest for conservation and sustainability through tourism.

## Refining the Vision of Ecotourists in Monte Verde, 1990s–2000s

Touring Monte Verde is primarily a visual activity, a search for things typically outside of normal human sight. During the past decade or so, several potent signs have dominated the grounds of what counts as worthwhile to view in the Monte Verde region and how to make sense of these views. This is in spite of two important interrelated facts: 1) the "tourist gaze" is differentiated and promiscuous, and therefore not reducible to any singular search or visionary practices (Urry 1992), and 2) there is a cultural diversity of visitors to Monte Verde, including a majority of North Americans (65 percent), Europeans (30 percent), and lesser numbers of Costa Ricans and other Latin Americans (5 percent) (Sanchez 2002). Nonetheless, for many visitors, touring Monte Verde means searching for and photographing certain of its constituent and iconographic aspects from different angles (Vivanco 2001). This includes visually "taking in the scenery," and searching for panoramic views of the distant Pacific lowlands, rain and cloud forest landscapes, and nearby volcanoes. With the development of so-called "canopy tour" attractions since the mid 1990s, it can imply either situating oneself above forests for a more comprehensive view of the landscape, or gaining an adventuresome new perspective on forest structure (see figure 7.1). Central to these practices is searching for and gazing upon key species, such as the famed resplendent quetzal, bellbirds, and whatever other forest creatures show themselves.

Another aspect of this is viewing Monte Verde landscapes as representatives of wider projections of idealized landscapes, such as quintessential or authentic jungle. For example, in preparing readers for what they will experience and look for at the Monteverde Cloud Forest Preserve, one tour guide book describes walking through a cloud forest in this way: "Palm trees and bamboos bent menacingly over the trail and I felt as if I were walking through a Grimm fairy tale" (Rachowiecki 2002: 214). Other narrativized mysteries help orient touristic vision-seeking, such as the disappearance of the golden toad in 1989, and so the hope is that one might be the first to see it again. But as I will show below, there is a tension between representing cloud forests as exemplars of universal phenomena ("Grimm fairy tale," mysterious landscapes) and as places with specific histories, which manifest themselves in the ways visitors come to know the forested landscapes themselves.

"Culture" and "community," insofar as they are understood as human inhabitants and their social relationships and traditions, are generally not promoted as worth viewing, although there are several exceptions. For the most part, Monte Verde ecotourism has tended to categorically disregard *campesinos,* or Costa Rican farmer inhabitants and their modes of life, as objects of gazing or interaction. One major reason for this is that they are not indigenous or "cultural" in the same way that indigenous peoples are seen to have culture, cultures that permit them to live in harmony with nature as "ecological noble savages." A Costa Rican hotel owner in Monte Verde confirms, "Look, we don't have culture here for people to see. The attraction we sell here is nature." One North American who spent several days in Santa Elena during the late 1990s expressed a relatively common point of view among tourists: "When you see these Ticos [Costa Ricans] that live here, they seem just like other poor Third World people who either rely on tourism or they cut the trees down. They don't seem to have much culture that makes them interesting." One exception of this tendency is that since 2001, tourists have been able to visit coffee farms producing organic coffee for the Santa Elena cooperative. But the goal of these tours is not so much to view traditional modes of rural life, much less the details of local farmer knowledge, as it is to understand how organic coffee that meets international standards (it is certified fair trade) is produced. The point is to see how Monte Verdeans are implementing globally standardized practices related to agricultural sustainability.

Except for the (usually North American and Northern European) "development tourists" and volunteers (Global Volunteers, Youth Challenge International, etc.) who deliberately seek out the needy to help them with their

*7.1. One of ecotourism's latest innovations, a bridge through the forest canopy (photo by author).*

development and environmental problems, rural Costa Ricans have therefore been marginal to touristic attention. But Monte Verde does have "exotic" people and cultural icons, most notably the Quakers, and to a lesser extent scientific researchers. Within the narratives of Monte Verde tourism (in the slide shows that are presented to visitors at hotels, the presentations of naturalist guides, and guidebooks), the very history of Monte Verde itself tends to begin with the Quakers, ignoring Costa Rican settlements in the area since at least the 1920s. If these pre-Quaker patterns of settlement are referred to at all, it is often in terms of a unidimensional story of "squatters" reducing forest cover, disregarding the complex patterns of land tenure on the rural Costa Rican frontier (Nygren 1993; Vivanco 1999).

The central imagery of the Quakers in these presentations and guidebooks is as peaceful and industrious pioneers, a romantic and persecuted ethnic group (several had served in jail for resisting the Korean War draft) that through hard work and foresight has established a prosperous community with positive social values (see Rachowiecki 2002: 197). This is an image that fits well with Costa Rica's image as a peaceful island within a conflict-ridden region (the very image that helped establish Costa Rica in the 1980s as a viable tourism destination). Also at the center of these images are narratives of Quaker land-use stewardship and wisdom: "the Quakers had respect for the principles of conservation from the beginning" (Caufield 1984/91: 116–17). Ironically, it was the Quakers' establishment of a dairy industry that motivated forest destruction in upland areas of Monte Verde during the 1960s and 1970s, as Costa Rican farmers sought to join the Quaker milk cooperative; or that the Quakers have been a consistently united community with stable philosophies of land stewardship. Yet, as one Canadian tourist marveled after attending a slide show on Monte Verde's history put on by one of the original Quaker settlers, "this place has such a high purpose," reflecting his own satisfaction at being a brief participant in this socially and politically progressive community.

Interestingly, in this imagery the Quakers do not fit into the landscape in the same ways that indigenous people often do in touristic contexts, that is, living in intimate or spiritual harmony with their environment. This is because, having rejected dominant North American values of militarism and industrialism, the Quakers have adopted a preservationist stewardship over the land by not living within or materially exploiting its resources. They did not evolve from "primitivity" into that stewardship, but from an enlightened modernist vantage point that brought them to reject forest destruction. Consequently, Monte Verde's history is often defined by guidebooks and the narratives of tourism guides in terms of the positive efforts of prescient Quakers and their international environmentalist allies to protect tropical nature. The effect of this narrative is not only to reinforce the notion that rural Costa Ricans have played a negative role in

Monte Verde's environmental history, but perhaps more importantly, to confirm for visitors that nature reproduces itself without reference to the social histories and political economies that constitute it (Williams 1980; Wilson 1992).

The point is not that these images of Quaker environmentalism are somehow inaccurate, even if they do reduce the internal complexities within the Quaker community. More importantly, these projections represent powerful narratives and stereotypes to orient touristic vision and circulation patterns, patterns that tour guides, guidebooks, and slide presentations confirm and reinforce. Ultimately, for the vast majority of the tens of thousands of annual visitors to Monte Verde, seeing a Quaker, much less interacting with one, is a remote possibility. There are several reasons for this, including the fact that the Quakers neither present themselves as touristic curios, nor are they stereotypically identifiable. One British visitor remarked, only half-jokingly, that she expected the Quakers to look like the "Quaker on the oatmeal box," and was a little disappointed that she could not pick them out. More importantly, however much ecotourists may want to interact with Quakers, culture and community are as marginal to ecotouristic practices as viewing tropical nature and certain species is *central* to them. In the next section, I will discuss how the space of tropical nature is prepared for consumption so that certain features of the landscape are emphasized to fit certain self-serving environmentalist narratives, and user movements are standardized to manage their impacts.

## Managing the Touristic Landscape

In recent years, the number and variety of Monte Verde tourist attractions has burgeoned to accommodate the almost quarter million annual visitors. These include a butterfly farm, a serpentarium, a frog house, a coffee roaster, innumerable gift shops (including one where hummingbirds are attracted in great numbers with nectar feeders), an insectarium, an orchid garden, horseback riding operations, and private forest reserves. These attractions often separate out a particular element of the tropical natural world and put it on display to entertain and educate. Cloud forest reserves continue to be central attractions, serving as the "anchors" much like large box stores might serve as anchor stores for strip malls. Whereas all these other attractions offer simulations or simplifications of tropical forests, forest preserves putatively offer the "real thing." This is a false dichotomy, however, because forest reserves are just as often carefully managed spaces, where certain practices of separation and simulation help guide visitors both physically and intellectually through the forest. As the the prominent cloud forest reserves in the Monte Verde region, both the MCFP and the RSE are managed in ways that reflect and confirm authoritative concepts of museumized and unpeopled wilderness, as well as the fact that nature reproduces itself independently of the people who live there. In his important work, Horne (1992) calls these processes "framing," or the means through which planners, managers, and operators communicate to tourists how to look at things, and how to revere those select things as they become isolated from the rest of the world (Hollinshead 1999: 270).

Although the administrators and workers at the RSE have communicated their efforts as uniquely Costa Rican, locally specific ways of conceiving of and interacting with forests are absent or marginalized as "folklore" in the placards of the visitor's center, if they are recognized at all. Aside from a belief held by some Monte Verde Costa Ricans (including the hotel owner quoted before) that their culture does not offer any unique insights to knowing the natural world (ignoring, for example, local uses of forest materials for healing, food, and construction), this also reflects the international work crews that helped Santa Elenans design and build the reserve in the early 1990s. It also reflects the desire of managers to standardize the uses of the space across diverse users, including tourists who come with varying expectations of engagement with the forest, volunteers who come to offer their labor, and researchers. Managerial goals at the RSE, as they are at the MCFP, are organized around reducing negative impacts on the landscape and visitor experiences, and in ensuring that tourist encounters with tropical nature both challenge users to see and experience new things, while helping them to grasp prominent environmentalist story lines.

Managers achieve these goals, for example, by constructing trails to emphasize certain features of the landscape (waterfalls, panoramic vistas of the lowlands or volcanoes, bird nests, etc.), and often in zigzag fashion to maximize feelings of relative solitude and reduce the potential that visitors will have their experience of the cloud forest marred by seeing people on the same trail. Realizing that visitor complaints rise when there are high numbers of people in the forest, the MCFP has limited the number of people allowed in its reserve at any one time to 120 since the early 1990s to avoid the possibility that visitors will see more humans than animals or birds.[1] Although RSE managers have considered limitations, they have never imposed them. Having said that, it is apparent that among visitors, it is more appropriate to see other tourists than non-tourist or reserve workers. According to one North American visitor to the RSE during the late 1990s, the fact that she saw several young Costa Rican men walking through the forest was disconcerting to her because she did not know what their intentions were, implying that they might be hunters or be otherwise engaging in destructive activities along the trails.

Trails are also constructed to reduce visitor impact on biophysical forest processes, for example, by using wooden and concrete pallets to reduce erosion. If such trail construction is commented upon at all by users, it is generally viewed as a positive intervention for the long-term sustainability of the trails, a notion that is introduced in visitor center placards that explain the practice. Interestingly, at the same time that they prevent erosion, concrete or wooden pathways through the forest implicitly demarcate the socially safe space of trails from the wilderness beyond, and reinforce a "look but do not touch" ethic. Tropical nature continues to exist "out there," a mysterious and complex Other beyond the trail's limits that many visitors would just as soon not transgress, and managers would not want them to trample. The language visitors employ here is not necessarily one of discussing the otherness of nature, but typically through expressions of fear of "getting lost," or the language of adventuresome experience.

Maintenance workers at the RSE recognize this as a primordial fear of visitors and regularly cut back the vegetation that threatens to engulf the trails. They also want to open strategic vistas for pleasant views and to reassure those walking the trail that the way ahead is clear. But maintenance workers also distinguish between trails and their users so that some trails are cleared more vigorously and regularly than others. According to one friend of mine, when he began working at the RSE in the mid 1990s he was informed by coworkers that the people who walk shorter trails tend to be on package tours or visiting the reserve only for a short time, and do not appreciate encroaching vegetation. The explanation they gave him is that these tourists are *más delicados* ("more delicate") and *menos aventureros* ("less adventuresome") than others, and are usually identified by their membership in a large tour group, or the relative practicality of their clothing and shoes. He was told that visitors who walk on the long trails tend to desire a more significant sense of adventure (they are usually younger and dressed for more rugged conditions), and so those trails should not be maintained as frequently. There are several potent dynamics here, one being of course that the "adventure" of visiting a cloud forest is profoundly mediated by managerial concerns that visitor activities be predictable, controlled, and adjusted to different levels of expectation and comfort, even if many visitors do not recognize it as such.[2] But another aspect of this is that tourists alone do not engage in the objectifications that orient ecotourism practices; RSE workers also project what tourists want, and not necessarily even based on personal interactions with them (usually because of language limitations), but on stereotypes based on tourist age, clothing, and so on.

Since many visitors who come to the preserves are there for only short periods (about three hours on average) and some of them are in guided groups as large as ten or even twenty, there is the potential that they will leave without having seen what many came expecting to see, especially iconographic birds and mammals such as jaguars, toucans, and monkeys. Managers recognize that visitors expect to view natural residents and interactions but will likely not because, as one RSE director pointed out, "this is not a zoo." It is in this context that the quetzal becomes a powerful touristic spectacle, a key symbol around which the experience of visiting the landscape is organized.

# The Quetzal Spectacle

When they are nesting in Monte Verde's cloud forests, the resplendent quetzals figure as a central exhibit of tropical nature and spectacle of consumption. In a context where animals are difficult to see through the intense profusion of botanical life that are cloud (and to some extent rain) forests, the male resplendent quetzal represents an unusually spectacular sight, with its long tail and emerald green and red colors (Vivanco 2001). It is not spectacular on universalizing aesthetic grounds, however (local Costa Ricans have only recently come to distinguish it from other forest birds, for example), but it is in a theoretical sense: exhibiting, viewing, photographing, and representing quetzals magically detaches them from the rest of the world (Horne ibid.) and sets them up as representatives of tropical nature's redemption by triumphant environmental activism.

Resplendent quetzals are migratory birds, and during the times of year that they are passing through and nesting in Monte Verde's forests and cattle pastures, it is common to find tourists asking for them and scrambling through forests and pastures to view them. The number of tourists demanding the experience of viewing a quetzal is not insubstantial, since quetzal nesting period coincides with tourism's high season (March-April). During this time, when several thousand tourists may be visiting each week, it is not unlikely to encounter crowds of people craning their necks to view a quetzal nest, the ultimate goal being to view the male, with its dramatically long tail feathers. Much like visitors to a theme park, visitors to Monte Verde and its forest preserves expect to view things that are outside of normal human sight, a condition the quetzal fits well because of its much-acclaimed beauty (Davis 1997: 97; Jufosky 1993). As touristic spectacles, viewing, recording, and representing quetzals carry significant moral connotations, because, as MacCannell (1989: 235) points out, "Spectacles, great and small, are productions of moral-aesthetic equations and ratios which provide evaluative frameworks for the full range of human qualities and accomplishments." But this activity carries differential significance and represents distinct dreams, opportunities, and responsibilities for participants in the Monte Verde tourism economy.

At their most spectacular, quetzals are paradoxically fixed in time and timeless. They represent the redemption of nature by the triumphant international efforts to save Monte Verde from the local population. The act of viewing a quetzal fixes it in a specific time and space as a sight to consume within a quest to register other sights (Berger 1980). In other words, the desire to view quetzals exists within a structured process of bringing nature under human management and purview, and reduces their viewing to one moment in that process. It is possible because the highly specific social, economic, and political conditions that created the possibility of the viewing are obscured by the intense visibility of the bird itself, reducing attention on the wider conditions of its production. At the same time, the quetzal represents timeless and transcendent Nature, inhabitant and representative of an Edenic tropical landscape unbroken by humans. One North American biologist and environmental activist invoked transcendent language when he told me, "The fact that there are quetzals here in Monte Verde is a reflection of healthy habitat, and the necessity of maintaining that habitat in its undisturbed state if those quetzals are going to remain here."

Environmental activism to save and touristic desires to view quetzals are redemptive and nostalgic in another, more mystical sense: they connect the modern with a premodern and sacred past. Naturalist guides often point out that the quetzal was sacred for the Ancient Mayas (cf. Jufosky ibid.). In this sense, searching for and viewing quetzals reaffirms the moral imperatives of forest preservation and pristine wilderness, or saving things as they have always been (or at least were when the Ancient Mayas were around), since these creatures would not survive without forest cover. Quetzals, in other words, represent a dream of authentic and undisturbed landscapes, and viewing them is participating in a form of time travel and nostalgic mysticism (perhaps even a "sacred journey"), referring to the time before the anthropogenic fragmentation of their habitat (cf. Graburn 1989; Urry 1992; Thomas 1993: 23).

Furthermore, the reconnection to a sacred past lays the groundwork for a re-sacralized present and future, except instead of imitating the Mayas, it is from the vantage point of the (perceived) rational management of natural resources. The result will be that unlike what happened to the Ancient Mayas (the current theory in environmentalist circles being that the Mayas destroyed their habitat because they overused their resource base), the future of modern humanity's relationship to the environment will be based on sustainability and a new cultural sensitivity toward landscapes like these, mediated by environmental organizations and their scientifically trained experts. If successful, there is a deliberate transcendence of time itself, a reconstituted timelessness for tropical nature. At the outset, it must be noted, this perspective is based on a dismissal of local Monte Verdean ways of knowing and relating to forests and social and economic histories, assuming (often inaccurately) that *campesinos* represent a unilateral tendency to deforest and destroy habitats, and therefore offer no practical methods or ethical foundations for forest preservation.

But spectacles are also especially vulnerable to disbelief, if spectacle is also understood as staged performance (MacCannell 1989). Environmental activists and most quetzal-seekers do not consider helping quetzals by saving tropical forests to be staging anything, and so the nature one encounters in a cloud forest preserve is no less believable than it would be if that nature were unbounded and unprotected. But for the landowners and hotel owners that deliberately create and maintain spaces where quetzals nest, and then guarantee their viewing to naturalist guides and their tourist clients (for a price), the accusation of fakery is a potential challenge to the naturalness of the encounter. It is also true with the common practices of placing hummingbird feeders at forest preserve entrances, or feeding animals such as coatimundis and planting certain flowers near forest entrances to attract butterflies and hummingbirds (making them what Davis 1997 calls "routine surprises"). It is done so that visitors can view these species in close and predictable proximity. Certainly for some visitors (especially those that believe that this artificialization of natural interactions displaces the real ecological interactions of the species) these are inauthentic practices and worthy of criticism. It is not uncommon to hear the comment of one North American woman who, as we witnessed other visitors feeding a coati at the entrance of the Monteverde Cloud Forest Preserve, said, "Isn't a shame that this animal is so reliant on humans? It really bothers me that they condone feeding the animals and birds. Now this creature will rely on humans for the rest of its life."

It is here in this woman's skepticism that the contradiction involved in the relationship between tourism and the environment is most poignant: the large-scale desire for certain unobstructed images and unfragmented landscapes profoundly, if not negatively, affects the landscape and its constituents that are the object of the gazing (Hill 1990; Zurick 1995; Urry 1995). This quandary is also not lost on many tourists, who often justify the contradiction in this way (as told to me by a North American student visiting Monte Verde for a month): "Ecotourism is killing what you love. Since so many people are trying to see these remote areas, there will be inevitable damage. But on the other hand, you want to see it before it's gone." This young man is expressing the ultimate triumph of a global (post)modernity based on consumer capitalism (of which he is self-consciously part) that is destined to destroy distinctive ecosystems like Monte Verde. Ecotourism is viewing nature before its inevitable demise, but in a sad and ironic twist, those who travel to consume it are ushering in its extinction because of their own negative impact on the species and landscape.

And yet the chance to view a quetzal for an unobstructed photograph is a powerful urge, even for the critics of artificially attracting and sustaining animal populations. It is related to the belief that underlying the artificiality of the viewing is a fundamentally transparent and credible natural process. One North American tourist who visited a dairy farm with a naturalist guide to view quetzals nesting in a box created specifically to attract both quetzals and tourists commented, "You see, their behavior is still natural. I got my pictures of them ... I think the best will be the ones where I waited until they went away from the box and landed somewhere else." This man's observation shares an underlying assumption with the commentaries above, which is that there is an essential separation between humanity and nature "out there," and the latter's authenticity lies in this separation. Except in this case, this man rationalized any intervention across that divide in terms of the fundamental independence of the natural processes he witnessed: they prefigure animal behavior so that any intervention is irrelevant. Practices of photography that include capturing the image when it is not in the setting of an anthropogenically established nesting box prove the ascendancy, independence, and authenticity of tropical nature itself.

Such encounters between visitors and quetzals are possible because Monte Verde hoteliers and the travel agencies that bring tourists to Monte Verde have helped guide touristic attention to the quetzal, mainly by participating in the marketing of Monte Verde as the privileged location to view them in Costa Rica. For many of these businesspeople, their relationships with quetzals are based not so much on nostalgia for certain landscapes or pasts in which tropical nature stands independently of modern humans, but on a future based on tangible economic prosperity through the use of quetzals to attract and channel tourist spending. For many of these people, the quetzal gains eminence in spite of (and often in tension with) already existing modes of relating to cloud forests that have privileged certain kinds of forest resources for their utility (such as certain kinds of wood or other materials for construction, or medicinal plants) and dismissed quetzals as nonuseful or irrelevant.

Monte Verde is not the only place to view quetzals in Costa Rica (nor is it the only cloud forest), but it does have the infrastructure to support around sixteen hundred visitors at any one time. Not only was one of Monteverde's first guesthouses (popular among North American and European bird enthusiasts) named Pensión Quetzal, but the image of the quetzal has also been incorporated in hotel signs and in the imagery and language of marketing brochures. For example, one hotel brochure in English characterizes the Monte Verde cloud forest as quintessential cloud forest and the quetzal as the quintessential Monte Verde attraction, sharing a similar language that many travel agencies and hotels use to describe Monte Verde: "In the misty cloud forest, visitors will see one of the richest and most abundant displays of flora in the world. The cloud forest houses in [sic] a wealth of wildlife. The area in which this [sic] forests are found is one of the few remaining zones in Central America where the resplendent quetzal finds protection. Sightings sometimes depend on patience and luck" (Hotel Heliconia Monteverde, n.d.). At a number of hotels, quetzals have become "routine surprises" (Davis ibid.) during certain times of the year: there are quetzal nests on the property so that guests can view them without having to go far from their rooms.

Another significant aspect of the spectacularization of quetzals is that they are presented and sold for tourist consumption by people who do not otherwise work directly in the tourism economy. Some farmers have done this by attracting nesting quetzals and charging naturalist guides a fee for letting tourists view them. In this context, quetzals represent the possibility of supplemental income for some agricultural households (although they have rarely been the reason for conserving forests on these private properties, except for those farmers working with the Monteverde Conservation League's Bosques en Fincas program), as well as the positive recognition of environmental activists and tourists that they are supporting conservation.

Artists and craft producers have also incorporated quetzals into their creative expression, representing another set of priorities and desires invested in quetzals and their imagery (Vivanco ibid.). Women in the Monte Verde region have entered the tourism economy by producing crafts and artistic items for sale to tourists that feature quetzals. Many market these goods through CASEM (Comisión de Artesanos de Santa Elena-Monteverde, or Artisans Commission of Santa Elena-Monteverde), an art and craft cooperative based in Monteverde village that is a popular place for visitors to buy souvenirs. CASEM grew out of a 1970s initiative of Monteverde Cheese Factory officials who believed it unhealthy that the Monte Verde economy rely so heavily on milk production. A Cheese Factory study on economic diversification found that many rural Costa Rican women produced *artesanía* [handicrafts] in their homes, and in the early 1980s a market for these handicrafts was established to sell to residents and the few tourists who visited Monte Verde. Participants in these craft bazaars, among them Costa Rican women and Quakers, decided during this period to formally affiliate with the local agricultural cooperative, Cooperativa Santa Elena, to provide their organization with a formal institutional and financial framework. CASEM's central goals have included providing employment for women, training members in diversified craft production and business management, and less formally, to provide women with opportunities for mutual support, creative outlet, and increasing self-confidence and decision-making ability (Leitinger 1997). As tourism, and the subsequent demand for souvenirs, has grown, CASEM's membership grew (in 2004 there are ninety-four member artisans), drawing from communities and hamlets throughout the Monte Verde highlands.

In many cases, women who have worked with CASEM have seen quetzals only in photographs of tourist guidebooks before they begin reproducing them in their *artesanía*. According to a middle-aged woman who embroiders images of quetzals on shirts and napkins,

> I have only seen a quetzal once, and that was because I really wanted to see one and a friend who is a naturalist guide showed me. This was very exciting to me since I had already been embroidering them. No one paid any attention to them here when I was younger, before the tourists began to show interest in them. They were just common birds to us. I have embroidered the quetzal based on a photograph that my husband got for me. I embroider other birds now, like the toucan, because I know tourists like them.

Monte Verde environmental activists have contended that such desire to aesthetically consume and reproduce quetzals is an indication that appreciation for quetzals is increasing among residents, and that without habitat preservation there would be no possibility of this appreciation. Therefore the value of the quetzal (if not intrinsic) is in its ability to induce residents to reconceptualize their relationship with forested landscapes and biodiversity and to provide the groundwork for the preservation into perpetuity of those forests. There is no doubt that there is a new appreciation of quetzals (and increasingly other touristically desirable creatures) among many Monte Verde residents. But quetzals do not simply offer an object for aesthetic representation to many women of Monte Verde, since they also represent opportunities for income and the dreams of present and future economic improvement. Indeed, one resident of Santa Elena told me she thinks quetzals are *muy linda* (very pretty), and then joked that her personal private image of the quetzal has its long tail reworked into a dollar sign ($)!

To many of the women who contribute *artesanía* to sell in CASEM, embroidering quetzals (and other forest birds) represents an opportunity to make their own contribution to family income and to increase their financial independence of their husbands and fathers, if not enhance their self-esteem and provide an outlet for their creativity. Indeed, this contribution has been the intention of CASEM founders and activists since the beginning of the organization. Furthermore, income from CASEM sales is widely believed to represent an essential component for basic subsistence in many of the households where these artisans live, contributing to a significant transformation in the gendered aspects of Monte Verde economic structure so that women are increasingly viewed as important economic agents.

So not only are quetzals and other images of touristic consumption centrally incorporated within intentions and possibilities for increasing (at least economic) gender equality in Monte Verde, but they are icons as well for the feminization of the Monte Verde craft and souvenir market. Initially, this process of feminization has contributed to tensions within families as men have had to recognize their wives' and daughters' increasing contributions to household income and economic independence. Stories abound among women involved in CASEM of the initial difficulty they had convincing their husbands to allow them to attend organizational meetings. It is important to note that this feminization process is not limited to craft production, since many women are now working as maids and cooks in hotels, so the lower levels of the tourism workforce more generally are experiencing feminization as well.

## Competing with the Spectacle: Narratives of Science and Environmentalism

Put in more theoretical terms, as spectacular displays of tropical nature and its redemption, resplendent quetzals are representatives of the impossibility of an unmediated encounter between the gazer and the object of gazing. Quetzals themselves mediate socioeconomic relationships between consumers and producers in Monte Verde's tourism economy, just as producers of tourism experiences mediate the relationship between tourist and quetzals to serve their ideological commitments to forest preservation or to benefit their immediate economic situations.

This impossibility for a neutral encounter is not mediated simply by the interests of capital and formal protection efforts, however, but by competing claims to knowledge as well. It is because quetzals must also be *explained*, not simply viewed. Indeed, it is wise to consider Taylor's (1994: 13) observation that "taking in a landscape was always (and often solely) a cerebral activity." In the context of Monte Verde ecotourism, it is not enough for tourists to view things because they must also intellectually connect what they see to what they already know, or are learning about, in the process of their movements through the concrete space of Monte Verde. A key framing of this is of course natural history and specialized scientific knowledge, and it is in this respect that seeing field-working scientists lends a sense of authenticity for certain tourists. In spite of the fact that touring the landscape is saturated with natural history knowledge, it is by itself not necessarily able to hold the attention of many tourists, even if most have a sense that it is an important part of their visit. As a result, broader stories of environmental transformation become even more important framing devices for making intellectual sense of Monte Verde.

Naturalist guides (there are currently some fifty who live and work in the Monte Verde area) often observe, usually in petulant but joking tones, that many tourists are consumed with the desire to view a quetzal. As interpreters of cloud and rain forest landscapes, a key concern of naturalist guides is to communicate specialist scientific knowledge to a lay public in ways that are tangible and meaningful. For guides and the tourists who hire them, Monte Verde landscapes are not simply a tropical "pleasure periphery" (Crick 1989) or setting of purely visual spectacles, but more importantly zones of intellectual curiosity. If tropical nature was transparent and had no secret knowledge to be revealed, there would be no reason to hire a naturalist guide.[3] And yet, there is a generalized understanding that it is impossible for a naturalist guide to reveal all these secrets in several hours' time, and so for guides, taking visitors through the forest is an (often stimulating) job of communicating basic knowledge of forest systems, opportunistically exploiting whatever fauna reveals itself, and responding to the particular needs and interests of the tourists.

According to one long-term North American resident who has guided visitors in Monte Verde forests since the 1980s, a persistent aspect of his job is a negotiation between his desire to impart certain knowledge with the desires and expectations of tourists themselves,

> People I guide are often asking me if they will see a quetzal, and I do feel pressure
> to show one to them, especially when they say that is what they really want to see.
> Sometimes you simply won't find quetzals, since they're not in the area, so you use

it as an opportunity to talk about their migrations and the importance of preserving habitat beyond Monte Verde. But so many tourists who come here think nature is like the posters you see in the gift shop, you know, the posters with all the animals and birds in the same frame.... And some days, or at some times of the day, you won't see any of these animals or birds you would like to show people. So you have to talk about other things, like the plants and their ecological interactions.

The educational stance is intended to demonstrate in a basic sense "how a cloud forest works," and is related to the relatively common perception among guides that "real ecotourists" do not come to Monte Verde as often as they once did. The concern is that tourism in Monte Verde is now a large-scale, casual, and mainstream phenomenon, and current visitors are generally not as knowledgeable or as interested in the natural history of the cloud forest as the "real ecotourists" who used to come. According to the same guide,

> When I think of ecotourism ... I think of the hardy people who weren't afraid to get wet and dirty and learn something about the forests.... Today the typical tourist we get here is different. They come to Monte Verde because they heard it was nice, or they've heard about the quetzal, or their tour group comes here. They want nicer accommodations and finer food. They don't know nearly as much about the forests, or have as much interest in environmental issues as the ecotourists of years ago. You can even see it in the way they walk, since they walk so fast when they are in the forest. In fact you could walk one hundred meters in three hours and see more than most tourists would see walking fast!

As MacCannell (1989: 115) has pointed out, one of the moral conditions of visiting a nature reserve is that a person leaves her social needs and desires behind to enter only with her essential humanity. The interesting fact of this insight is that of course no one ever actually does. But people do interact differently with forests, and guides will often differentiate between the North Americans who enter a forest somberly, as if it were a cathedral, and Costa Ricans, who often enter more boisterously, as if it were a familiar space of recreation. Nevertheless, naturalist guides often take advantage of this dictum, and their goal is to reframe the ways tourists make intellectual sense of the landscapes they visit, without really drawing attention to that process. In doing so, they often invoke overall grand images and broader messages about the forests, leading them to apply certain commonly accepted story lines about Monte Verde's environmental history that reconfirm tropical nature's redemption through environmental activism and ecotourism. I found the words of one young Costa Rican guide (to a group of Canadians) relatively common: "During the generations of our fathers and grandfathers, all they did was cut forests. But then the scientists came and with the Quakers they began to protect forests, so you see now we fortunately have a lot of forest." Tourists are also often told that their ecotourism is the last link in this chain of preservation, and the guides can often offer up themselves as examples of how successful ecotourism has been: it has helped them learn about and appreciate the biodiverse forests near which they live. As moralistic tales of human destruction of tropical nature, and its redemption through positive environmentalist intervention and the rise of ecotourism, these stories nevertheless often take on a linear quality of humanity's progress, so that more specific stories and dilemmas of nature, community, and environmental conservation are marginalized. These alternatives include more complicated and unsettling narratives, such as the fact that many Monte Verde landowners have not engaged in the same high levels of deforestation found in neighboring zones of the country where political-economic conditions differ, and in fact have maintained important amounts of forest on their lands, or the controversies sparked by the concentration of lands as forest preserves.

## Conclusion

Ecotourism, like other forms of adventure travel, draws from and serves as a conduit for an increasingly transnational concern with the disappearance of distinctive places, cultures and ecosystems (Johnston 1990; Zurick ibid.). Proponents argue that at its best, ecotourism insists upon a broad reconceptualization of the position of humans in natural landscapes, so that natural resources are not overtaxed, local people are offered economic alternatives to forest destruction, and perhaps most importantly of all, individual ecotourists are provided with spiritual renewal and a knowledge base of the natural world that will reinforce a sense of responsibility for the world's remaining wildernesses (Whelan ibid.; Budowski 1992). It is important to consider, however, how ideologies and practices of ecotourism themselves represent agents in extending certain universalizing narratives of environmentalist history and science while in fact contributing to the loss of distinctive places, cultures, and ecosystems. In other words, as Zurick (ibid.: 2) states, so-called alternative tourism carries with it "the very defilements that ... tourists wish to escape," because, among other things, it contributes to the perceived homogenization of the diverse cultures and landscapes that are deliberately sought by alternative tourism. Cosmopolitan tourists and tourism promoters do this by seeking out the pristine and untouched (not knowing, or preferring not to consider, their role in its social construction and objectification), and despoiling it by sheer numbers and the effects of their consumption patterns on local nature and waterways. Zurick's observation, while simplifying the interaction between hosts and guests to a one-way transferal of knowledge and culture, nevertheless forces us to consider if alternative forms of tourism like ecotourism are indeed alternative at all.

This issue is not simply a question of scale, although some Monte Verdeans have expressed concern over the numbers of tourists that visit them, the population explosion related to tourism development, and the impacts these factors have on regional ecology and society. It is because the lines of "adventure" have also begun to blur, so that the sensibilities of "soft" adventure in ecotourism are increasingly competing with the activities of "hard" adventure, the latter typically defined as the adrenaline-hyped attraction of capitalism's new generation of nature-based products: zip-lines, Tarzan swings, paintball battle experiences, all-terrain vehicle rentals, and so on. Monte Verde is not alone here, as these attractions, including some that have not yet reached Monte Verde (such as *teleféricos*, or skilifts over forests), spread throughout the Costa Rican countryside. This proliferation of offerings has meant that Monte Verde is appealing to new market segments beyond the idealized traditional ecotourist searching for unique flora and fauna. And the fact that many of these newer attractions are based in Santa Elena has motivated a shift in touristic attention and movement patterns, not so much away from Monteverde village where there still are important attractions, but into new rural spaces in the region that did not receive tourists a decade ago, expanding the footprint of the touristic economy even more.

I find it significant that in spite of an opportunity to construct new understandings of life and forests in rural Costa Rica, Monte Verde's production of landscapes continues to be projected and widely viewed in simplistic terms that stereotype and homogenize certain local people as ignorant and destructive and others as enlightened and morally superior. Or that there is no discussion in touristic spaces about the scale of tourism or the current shift toward high-cost and infrastructure-intensive constructions like canopy bridges in fragile ecosystems. Ultimately, these simplifications reduce the complex human and natural histories that have constituted the current landscape in the first place. The question to explore now involves the identification of how and where story lines alternative to these could be generated and authorized that reconsider unanswered environmentalist and tourism quandaries. These quandaries include, for example, how the "environmentally virtuous" have sought to consolidate their status as arbiters of Monte Verde's future by mystifying not only local cultural ecologies, but their own lack of appreciation for culturally different ways of interacting with Monte Verde's forests. Another, even more profound quandary is how the interests of nature conservation and tourism might better serve the interests of more broadly defined social equity, as raised in the panel in which I participated. Or how ecotourists can understand both the positive and negative effects of their activities on local ecology.

It has always seemed to me that an appropriate space to engage in such self-reflection is a consciously "Costa Rican" context like the RSE. It does not have to involve any essentialization of what is Costa Rican, for the grounds for defining what is Costa Rican are dynamic, partly the result of Monte Verde's very engagement with tourism and environmentalism themselves. But so far, this has not happened. Nevertheless, it seems relevant to continue questioning why in the context of Monte Verde ecotourism it is possible to casually ignore or dismiss complex stories such as the one with which this chapter opens, in which ecotourists avoided or ignored significant political debates over how the RSE would be managed. Indeed, it is these very politics that are constituted by and constitute local meanings of nature and community that in turn will have an effect on the long-term sustainability of the natural resource base and the people who live there. I believe part of the answer lies in the extent to which ecotourists and ecotourism promoters rarely, if ever, question the very epistemological foundations that they purport to challenge: ecotourism does not unite nature with community because it continues to see them as independent, not mutually interdependent, others.

## Notes

1. During the mid 1990s the MCFP also sponsored studies of tourist impact on wildlife populations, especially birds, in order to determine if visitors had a negative impact on attracting wildlife.
2. In my experience, many visitors do not express awareness of how the space is managed except when it challenges their own sense of comfort.
3. At the RSE, among tourists who arrive independently of organized package tours (which arrive with guides) between 20 percent and 30 percent hire naturalist guides for several-hour walks. During the tourism high season, there may be as many as fifty Monte Verde region residents who work as naturalist guides, which has become a prestigious and well-paying job.

# 8  Conclusion: Environmentalism at a Crossroads

One of the easiest and most seductive ways to think about Monte Verde's encounter with environmentalism is in terms of its achievements and successes. They typically include, as Burlingame (2000: 372–3) observes, the sheer amount of formally protected lands (somewhere around twenty-nine thousand hectares in total); the use of local residents as unarmed guards; the creation of windbreaks and biological corridors to support birds and farmers; widespread environmental education for youths; the emergence of local organizations to meet new needs; the substantial contributions that international scientists and fundraisers have made; and the rise of ecotourism as a way to financially support protection efforts. Ecotourism in Monte Verde has thoroughly incorporated these examples into its narratives, so moving through Monte Verde tourist spaces tends to reinforce the idea that effective alliances can be created between local people and international activists to solve problems of environmental degradation. After a few days' visit, enchanted by the rugged beauty of the place, one could easily drive away from it with a sense that Monte Verde justifiably occupies one of the top positions in the pantheon of the world's most eco-friendly places.

Friends of mine in Monte Verde who pay attention to these things call this lingering, even magical effect on visitors "the honeymoon phase." It should be said that this is not unique to Monte Verde at all. The very logic of tourism is to take facts about the world at large, decontextualize and reduce them, and rearrange them in ways that confirm the self-serving messages of those who created the attraction (Kirshenblatt-Gimblett ibid.). If a tourist walks away disillusioned, it is often because this process has revealed itself as nothing more than a ploy to take money from the tourist's pocket. Environmentalist discourses can have a similar tendency to reduce and rearrange facts about nature and the people who interact with it in ways that address the specialness and inevitability of environmentalism itself. Because environmentalism exists in a realm of evidence-based argumentation, this is often expressed in quantitative terms: how many hectares are saved? How many children have received environmental education? How many dollars do ecotourists provide for protection efforts? How much carbon fixation potential do protected forests hold to combat global warming?

So why is it that when I run across such statistics, quantitative indicators, and calculations, I feel a bit like that imaginary tourist above, that someone has their hands in my pocket? Let me qualify this question. I am not asking it because I am trying to disqualify these accomplishments or to dismiss them as irrelevant or wrong. In fact, what has happened in Monte Verde is profoundly important, and worthy of our respect, critical attention, and willingness for dialogue. I am also not asking it because I agree with right-wing critics who have recently been arguing that shrill environmentalists are perpetuating a fraud on the public with hyberbolic claims about impending global ecological collapse (Lomborg 2001; Rowell 1996). There is little doubt that narratives of ecological crisis, even those used to justify land purchase campaigns in Monte Verde, have sometimes framed the problem in self-serving terms that orient solutions toward preconceived outcomes. This issue raises significant questions about the comparative abilities of environmental activists to promote certain versions of reality in transnational media and fundraising contexts, and the comparative inabilities of people living there to counter or complicate those narratives. But there is little sense in debating the connections between deforestation, soil erosion, and drought, or the fact that increasing pollution and toxic emissions affect people's health and quality of life. The ulterior motive of so-called "right wing environmentalists" is not to look seriously at what causes and who pays the price of environmental degradation; it is a cynical effort to reorient debates about the environment to justify even greater market control over its exploitation.

One reason I ask this question is because of my discomfort with sound-bite evaluations of environmentalism that describe it in terms of quantitative successes (or even failures) alone. Telling us that environmentalism has been successful in Monte Verde because a couple of organizations bought a lot of land simply tells us that a couple of organizations bought a lot of land. It does not tell us how that was achieved because of profound and ongoing political and economic transformations; how and why it used certain self-serving projections of nature and *campesino* destructiveness; what the long-term biological effects might be of creating an island of biodiversity and why some species might be more negatively affected by the creation of protected areas than others; or if resentful and landless people might someday stage a land invasion. Telling us that foreign ecotourists are helping the MCFP pay to protect its preserve does not tell us how the MCFP would survive if tourists simply decided to go elsewhere; why an organization like the Monteverde Conservation League has not pursued ecotourism very seriously to pay for its forests; or how ecotourism narratives celebrate unpeopled nature and marginalize more complicated stories about human relations with the landscape.

Such declarations of success often tell us more about the unexamined assumptions environmentalists and other observers make than about the worlds in which they operate. This includes unexamined assumptions about rural Latin Americans (as passive and homogeneous subjects), environmentalists and their priorities (as united on a morally enlightened mission to save nature and humanity), tropical nature (as a static and Edenic wilderness), and how to measure that success or failure (as a relatively straightforward presentation of sustainability indicators). It is also common to assume that environmentalism is a positive product of Euro-American elite and middle-class culture and politics, and that it is spreading through the world in basically uniform and unproblematic ways. It is as if becoming involved in environmentalism is like assuming a new state of modernity, and that a few quantitative indicators can provide a definitive demonstration of this. It bears noting that in some very important ways, Monte Verde's encounters with environmentalism *are* in fact products of Northern elite and middle-class interests, especially if you consider the hundreds of thousands of people in industrialized countries who helped "save" Monte Verde by sending money to adopt acres of primary forest. Likewise, North American and European researchers, volunteers, fundraisers, and activists have played crucial roles in helping create, implement, and sustain programs in forest protection, biological corridors, environmental education, integrated conservation and development projects, reforestation, and so on. As a result of these efforts, many rural Costa Ricans are changing how they think about and relate to the landscapes in which they live.

Nevertheless, environmentalism is not reducible to a set of beliefs and practices unilaterally exported from First World "center" to Third World "periphery." One of the central arguments of this book has been that Monte Verdeans' engagement with environmentalism has been uneven and dynamic, reflecting the fact that there are multiple visions and practices of environmentalism operating in a scene of complicated regional social, economic, political, and ecological change. The example that opens this book— of Manuel viewing a quetzal through a camera instead of through the sights of a rifle—is richly indicative of this unevenness. Or consider the situation that unfolded on Solórzano's *finca* in Chapter Four, in which he chose to negotiate over the life of the puma instead of unceremoniously killing it. In both cases, these men's apparent conversion to eco-sensitivity is exactly the kind of behavior that environmentalism seeks to engender in local people. But as I mentioned, on occasion Manuel still hunted, and Solórzano did not give his lands over to create a puma sanctuary, much less say he would never again shoot a puma. Similarly, the fact that many residents see the now-protected forests as off-limits to their recreation and use reinforces the authority of the environmental organizations that police those lands, but fuels quiet talk by some people of future land invasions, not to mention feelings of disconnection and even hostility toward the outsiders and foreigners who seem to now be enjoying the landscape. In yet another case, the people of Santa Elena may be talking about their "community" ecotourism and conservation project, but defining who counts as members of the community can generate sharp conflict.

Of course, I recognize that success and failure represent the central criteria upon which environmental activists and natural resource managers, their international and national sponsors, and the targets of environmentalist initiatives anticipate, evaluate, and justify outcomes. Whether or not an integrated conservation and development project like the San Gerardo Project is successful is extremely important for the people who lived there, whose daily lives were altered by the project long before it was dissolved. It is also important for the credibility of the non-governmental organization that became involved, and the government agency that invited them in. If a project can be shown to be successful, donors and governments are more likely to send money, technical experts are more likely to offer technical assistance, residents are more likely to support its programs, people begin to refer to it as a "model" for other places, and so on. On the other hand, a project's failure could mean decreased access to funding and a loss of prestige. Widely circulating notions like sustainability and participation appear to offer the grounds upon which to measure success and failure, albeit in nonspecific and relative ways.

But while it may seem to offer an objective, even technical, set of criteria for evaluation, we have to remind ourselves that outcome-based language is also a political language that renders certain things visible and others hidden. As a result, understanding the outcomes and instrumental effects of something like the San Gerardo Project is by no means a simple or straightforward exercise. It is because criteria of success and failure vary across and within institutions, communities, and times, and it is necessary to identify and contextualize who declares either and why they do it. In San Gerardo Project (to continue with this example), some people from San Gerardo and the MCL consider it to be an unmitigated failure, although for diverse reasons. For some, especially those who wanted to return to live "sustainably" in San Gerardo, it demonstrated the fact that the MCL had no real commitment to the original goals of the project. For others, it was the people of San Gerardo themselves who were unable to pull it together. On the other hand, some have considered San Gerardo successful, and on equally diverse grounds. They include arguments that the landscape is now protected in perpetuity, or that a biological station with ecotourism potential was built, or that people learned from the experience and will try to not make mistakes again. The debating point here is that in both cases, describing and analyzing the San Gerardo Project is constituted as an unproblematic choice between two limited alternatives. That is, the choices are rigged ahead of time so that we have to choose between one outcome or its opposite: yes or no, sustainability or nonsustainability, success or failure. When virtually everyone claims to want success and sustainability (is it not politically correct to reject these?), it becomes more difficult to raise critical or uncomfortable questions about processes: on whose terms are success and sustainability defined and why, for whom do they apply, and at whose or what expense?

I once had a marathon eight-hour interview with a former MCL reforestation official in which he said something that bears on this issue. He told me, "One of the sadnesses of Monte Verde is that people have different necessities, but environmental organizations and activists have never fully recognized this. Conservation has had the power to purchase lands, buy posts to make fences, and so on, but it has never bothered to know the priorities of rural people." It is a polemical judgement, asserted in the context of a long conversation about the possibilities of an environmental activism less committed to abstractions like the number of hectares of wilderness it protected, and more committed to the concrete realities of people living on the edge of existence in increasingly degraded rural landscapes. I have no doubt that many people featured in this book disagree with this, on the grounds that reforestation and other social work programs have been oriented toward improving the long-term livelihood prospects of rural people. Furthermore, a number of rural Costa Ricans I know themselves consider the work of environmentalism in formally protecting forest and expanding a tourism industry to be the most positive force in Monte Verde's history. In that sense, this individual was being unfairly essentializing, since some of those very same "rural people" to whom he was referring have asserted their priorities within environmental institutions, and these in turn have sometimes responded favorably to such interventions.

But it is highly relevant that the person who said this was a key member of one of Monte Verde environmentalism's most successful initiatives. His work in reforestation pleased farmers, who improved their milk and coffee production; biologists and other technical personnel, who wanted to see more native species and erosion control; and tourists, who have appreciated more trees in the pastoral landscape. He makes a provocative point, one that I have visited several times in this book, which is that in certain crucial instances, especially those in which international funding was at stake, rural Costa Ricans tended to represent subjects to be acted upon, and only collaborated with insofar as it helps achieve certain institutional, strategic, or ideological goals. In other words, environmental activists and institutions have often assumed that they know what is best for rural Costa Ricans, but have only rarely approached them on their own terms to see if what they need is what environmentalists have to offer.

And of course, as I have shown in this book, rural Costa Ricans have not responded passively to this situation. The point is not to pretend to know with certainty, much less to reduce to simple language, what rural Costa Ricans' priorities are (for, as the man quoted above insisted, they are multiple), as it is to challenge the presumption that environmental activists and institutions always know what those priorities are, to expose the ways they depoliticize such issues, and to multiply the possibilities for constructing locally responsive and transformative environmentalisms.

It is common in environmentalist discourse to advertise and vilify multinational corporations and capitalist interests as enemies of biodiversity. In spite of its potential costs (expensive lawsuits and public relations wars), these struggles have an aura of heroism, because of their David and Goliath quality. They also reflect an image of a politically progressive agenda for environmentalism that many middle-class North Americans and Europeans find comfortable, based on a critique of the antidemocratic tendencies and lack of accountability of multinational corporations. But it is more problematic when environmentalists vilify the people who live in and near the wildernesses they desire to save, or to charge that environmentalists can act in these circumstances in antidemocratic ways with little accountability to local people. Ironically, the people who are being alienated when local concerns are ignored are the very people environmentalism seeks to "turn green" in order to save the world's last wildernesses. Even more importantly, as the former reforestation officer suggested in my conversation with him, saving the rain forest may be a worthy cause, but it has not necessarily coordinated with the political progressivism of activists like himself who have been fighting to keep issues of rural poverty, structural inequalities, and the unbalanced distribution of resources on national political agendas. For him, ecological degradation is directly related to inequalities and social injustice, where poor people have historically been forced to live on marginally productive lands in the highlands as the productive lowlands have been captured by national elites seeking to produce for export markets. It is thus no surprise that people should view the concentration of lands as nature preserves "for the benefit of humanity" in the highlands as compounding historical problems of structural inequality. In fact, demonstrating that no institution is monolithic, the MCL did have (internally controversial) spaces where such issues and processes were considered and addressed, through the reforestation program and, in its early years, the environmental education program.

Some cosmopolitan analysts have begun to assert that preservationism's dominance in the environmental movement has focused on "scenic nature and cuddly animals" at the expense of incorporating broader agendas of social transformation, such as occupational health, social and political justice, and nonwhite and working-class diversity (Dowie ibid.; Gottlieb 1993; Guha and

Martinez-Alier 1997). It has led at least one critic to assert that environmentalism as a force for positive social and natural transformation has been "courting irrelevance," as demonstrated by a general decline in contributions to mainstream U.S. environmental organizations, as well as the reluctant incorporation of environmental problems of poor and minority communities (Dowie ibid.). There has also been a brewing backlash against U.S. nature protectionist agendas (Rowell ibid.), a hostility that has been emanating from the highest levels of government during the Bush years. But while wilderness preservation may be "courting irrelevance" here in the U.S., this is not the case for people living in places like Monte Verde. In Costa Rica, it is difficult to accept that conservation is courting irrelevance, if recent developments are any indication of the extent to which that country's political culture and tourism-based economy have incorporated ideologies and legislation for nature protection, even as corruption, mismanagement, and deforestation persist. These include additions to the national park system during the past decade, the continuing centrality of nature-based tourism to the national economy, the increasing interest of multinational corporations in tropical landscapes as reservoirs of biodiversity and biotechnology contracts, and legislative projects to protect biodiversity and wildlife. It has been and will certainly continue to be argued that Costa Ricans are not "conservation-minded enough" (J. Hunter 1994). And there is certainly less international money available today for Monte Verde environmental organizations than there once was, seriously limiting their abilities to keep certain programs alive. But this should not obscure the fact that there is an ongoing redefinition of the Costa Rican countryside as a place of nature, and a new ordering of rural communities as buffer zones where there is increased scrutiny on how people relate to the landscape.

Like environmentalism itself, Monte Verde is also at a crossroads, and it is clear that these processes are intertwined with each other. The explosive growth and scale of Monte Verde's nature-based tourism economy is clearly an effect of environmentalism's successes. To have almost a quarter million annual visitors in an area whose population is four thousand (and still growing, it appears) presents substantial business opportunities for some (and jobs for many others), while posing serious dilemmas for everybody. These problems range from highly contaminated local waters and uncontrolled urbanization to perceptions of increasing socioeconomic inequality and the social inconveniences of just getting around during tourist high season. They are especially acute given the lack of planning for urban and tourism development, and organizations such as the MCFP, the RSE, and the MCL have not dedicated much effort to these issues from an environmental point of view.

It is not clear if (or when) Monte Verde may reach a tipping point when tourists decide to go elsewhere instead of putting up with the increasingly aesthetically unappealing villages of Santa Elena and Cerro Plano, not to mention the sheer numbers of other tourists competing for hotel rooms, restaurant tables, and unobstructed views of quetzals and other forest creatures. A number of self-described ecotourists have told me in recent years that they now prefer San Gerardo de Dota, an area south of the Costa Rican capital that is widely known as what "Monte Verde was like thirty years ago," with similar cloud forests, quetzals, and rustic accommodations. I suspect that Dota's current tourism mini-boom of its own is based only partially on people who have sought it out as an alternative to Monte Verde, and it is necessary to point out that Monte Verde's ecotouristic cluster continues to grow, not decline. But a place like Dota cannot help but benefit from a struggle over the soul of tourism and its relation to sustainability—should it be large or small scale, should it pursue quiet contemplation and learning or Tarzan swings—in which Monte Verde is increasingly known for its "mass" appeal and high-adrenaline adventure offerings. It seems relevant that in February 2004, the municipality of Monteverde gained a "sister city," Estes Park, Colorado, which is a town of five thousand on the border of Rocky Mountain National Park that receives three million tourists a year. It is a suggestive relationship, at least from the Monte Verde side of things, in that it implies something about how leaders perceive Monte Verde's equals and future: large-scale nature-based tourism destinations on the edge of wilderness.

The current economic dependency on tourism raises questions about not only ecological and social sustainability, but also the changing nature of Monte Verdeans' social and political autonomy. The productivity of Monte Verde's tourism sector, and the newfound political influence of some of its entrepreneurs, helped Monte Verdeans gain municipal status in 2001. The ambiguities of this situation are indicated in the increasing use and acceptance of the one-worded "Monteverde," which was normalized with the naming of the municipality. On the one hand, Monte Verdeans (North American and Costa Rican) from Santa Elena, San Luis, Cerro Plano, and Monteverde village, not to mention other outlying villages, have long worked together on projects of common concern and interest: road maintenance, the construction and maintenance of water canals, and conservation and tourism organizations. They continue working together now in the "muni" and its committees, although they now have a locally elected mayor, collect taxes locally, and no longer have to travel to Puntarenas for basic bureaucratic matters. On the other hand, the "muni" brings new bureaucratic presence and surveillance in an area that was typically marginalized by distant administrative centers. One of these is a process currently under way, the creation of a *plan regulador* (regulatory and zoning plan) for the lands and communities that fall within the municipality. This process has already generated significant controversy, as it brings central control over infrastructural developments, requiring environmental impact analyses and permits, something that few Monte Verdeans have really cared about. The fact that some people seem to get special treatment and access at the mayor's office fuels increasingly harsh political divisiveness and accusations of cronyism.

An instrumental effect of establishing the municipality is that MINAE, the Ministry of Environment and Energy, has opened a part-time office in Santa Elena, to assist in the enforcement of environmental regulations. It is the first sustained presence of the state environmental ministry in Monte Verde, although there have been long-term collaborations between MINAE forest guards and the forest guards of Monte Verde's environmental organizations. It has happened during a moment in the history of Monte Verde environmentalism when the once-dominant MCL has been in a multiyear decline because of the drop in international funding since the mid 1990s. MINAE's new presence has significance for local environmental politics, because of the political vacuum left by the MCL's decline. It is easy to overplay the significance of MINAE's presence, because even MINAE officials admit that they are stretched so thin that their ability to offer a consistent presence is limited, but the fact is, MINAE brings a new kind of authority (backed by coercive force, not persuasion) as well as resources to enforce environmental laws and regulations.

As the region confronts new realities, environmentalism in Monte Verde is itself currently going through a political reorganization. Such change is not all new, and as I have shown throughout this book, environmentalism has always been a dynamic and contested arena through which Monte Verdeans have encountered ongoing political, economic, and social changes at all levels—local, regional, national, and global. Throughout this history, environmentalism has been a channel for asserting new kinds of control and interventions in rural people's lives, but at the same time it has provided a space for vigorous discussion about alternatives to current patterns of development, including those provoked by environmentalism itself. The fact is that even while in their daily lives many Monte Verdeans currently seem more preoccupied with the activity of tourism, and the rancorous debates over controversial issues like land purchases have faded into the past, the cultural and political authority of environmentalism to set aside large areas of wilderness, attract tourists, and educate children is so well-established that most Monte Verdeans openly declare with pride that they live in a region with deep respect for biodiversity and sustainability. There are important contours to this, as this book has shown, but that these qualities are so central to how people define themselves should tell us something about how powerful and transformative Monte Verde's encounters with environmentalism have been.

# Bibliography

Acuña, Marvin and Daniel Villalobos. 2001. "Ecoturismo en Costa Rica: Competitividad y Sostenbilidad." *Ambientico* 98: 7–10.

Adams, Jonathan S. and Thomas O. McShane. 1992. *The Myth of Wild Africa: Conservation Without Illusion.* Berkeley: University of California Press.

Adams, Vincanne. 1996. *Tigers of the Snow and Other Virtual Sherpas: An Ethnography of Himalayan Encounters.* Princeton: Princeton University Press.

Adams, W.M. 1990. *Green Development: Environment and Sustainability in the Third World.* London: Routledge.

Agarwal, Arun. 2005. "Environmentality: Community, Intimate Government, and the Making of Environmental Subjects in Kumaon, India." *Current Anthropology* Vol. 46, no. 2 (April 2005): 161–90.

Album Committee. 2001. *Monteverde Jubilee Family Album.* Monteverde, Costa Rica: Asociación de Amigos de Monteverde.

Allen, W.H. 1988. "Biocultural Restoration of a Tropical Forest." *BioScience* 38(3): 156–61.

Alvarez, Sonia, et al. 1998. "Introduction: The Cultural and Political in Latin American Social Movements." In *Cultures of Politics/Politics of Cultures: Re-Visioning Latin American Social Movements,* edited by Sonia Alvarez, E. Dagnino, and Arturo Escobar. Boulder: Westview Press.

Anderson, David and Eeva Berglund, eds. 2003. *Ethnographies of Conservation: Environmentalism and the Distribution of Privilege.* New York: Berghahn Books.

Anger, Dorothy. 1989. "'No queremos el refugio:' Conservation and community in Costa Rica." *Alternatives* 16(3): 18–22.

Annis, Sheldon. 1988. "Can Small-Scale Development be Large-Scale Policy?" In *Direct to the Poor: Grassroots Development in Latin America,* ed. Sheldon Annis and Peter Hakim. Boulder: Reinner.

Appadurai, Arjun. 1999. *Modernity at Large: Cultural Dimensions of Globalization.* Minneapolis: University of Minnesota Press.

Appfel-Marglin, Frederique and Stephen Marglin, eds. 1990. *Dominating Knowledge: Development, Culture and Resistance.* Oxford: Clarendon Press.

Aspinall, William, et al. 1991. "Master Plan for Monteverde Cloud Forest Preserve." San José, Costa Rica: Tropical Science Center.

Augelli, John. 1987. "Costa Rica's Frontier Legacy." *The Geographical Review* 77(1): 1–16.

Austin, P. 1990. "Sharon Kinsman: The International Children's Rainforest." *E Magazine: The Environmental Magazine.* Vol. 1, no. 3 (May/June 1990): 47.

Aylward, Bruce, et al. 1993. "The Economic Value of Species Information and its Role in Biodiversity Conservation: Costa Rica's National Biodiversity Institute." London: London Environmental Economics Centre Discussion Paper, DP 93–06.

Aylward, Bruce, et al. 1996. "Sustainable Ecotourism in Costa Rica: the Monteverde Cloud Forest Preserve." *Biodiversity and Conservation* 5: 315–43.

Báez, Ana. 1996. "Learning from Experience in the Monteverde Cloud Forest, Costa Rica." In *People and Tourism in Fragile Environments,* ed. M.F. Price. Chichester: John Wiley & Sons.

—— 2001. "Turismo Nacional ante la Crisis Mundial." *Ambientico* 98 (Nov. 2001): 11–12.

Báez, Ana and Fernando Valverde. 1999. "Claves Para el Exito de Proyectos Ecoturísticos con Participación Comunitaria. El Caso Costarricense del Sky Walk-Sky Trek." *Ciencias Ambientales* 17: 18–24.

Bailey, F.G. 1996. "Cultural Performance, Authenticity and Second Nature." In *The Politics of Cultural Performance,* eds. David Parkin, Lionel Caplan, and Humphrey Fisher. Providence: Berghahn Books.

Bandy, Joe. 1996. "Managing the Other of Nature: Sustainability, Spectacle, and Global Regimes of Capital in Ecotourism." *Public Culture* 8: 539–66.

Barham, Elizabeth. 2001. "Ecological Boundaries as Community Boundaries: The Politics of Watersheds." *Society and Natural Resources* 12: 181–91.

Barlett, Peggy. 1982. *Agricultural Choice and Change: Decision Making in a Costa Rican Community.* New Brunswick, NJ: Rutgers University Press.

Barry, Steven J. 2004. "Parks in State of Emergency." *Tico Times* (March 26, 2004): 1 and 5.

Bedoya, Eduardo and Lorien Klein. 1996. "Forty Years of Political Ecology in the Peruvian Upper Forest: The Case of the Upper Huallaga." In *Tropical Deforestation: The Human Dimension,* eds. Leslie Sponsel, Thomas Headland, and Robert Bailey. New York: Columbia University Press.

Belsky, Jill. 1999. "Misrepresenting Communities: The Politics of Community-Based Rural Ecotourism in Gales Point Manatee, Belize." *Rural Sociology* 64: 641–66.

Berger, John. 1980. "Why Look at Animals?" In *About Looking.* New York: Pantheon Books.

Blaikie, Peter and Harold Brookfield. 1987. "Defining and Debating the Problem." In *Land Degradation and Society,* eds. Peter Blaikie and Harold Brookfield. London: Methuen.

Boll, John. 2000. "San Gerardo: An Experiment in Sustainable Development." In *Monteverde: Ecology and Conservation of a Tropical Cloud Forest,* eds. Nalini Nadkarni and Nat Wheelwright. Oxford: Oxford University Press.

Bonilla, Alexander. 1985. *Situación Ambiental de Costa Rica.* San José, Costa Rica: Ministerio de Cultura, Juventud y Deportes.

—— 1988. *Crisis Ecológica en América Central.* San José, Costa Rica: Ediciones Guayacán.

Bonner, Raymond. 1993. *At the Hand of Man: Peril and Hope for Africa's Wildlife.* New York: Vintage.

Boo, Elizabeth. 1990. *Ecotourism: The Potentials and Pitfalls.* Vol. 2. Washington, DC: World Wildlife Fund—U.S.

Booth, John. 1989. "Costa Rica: The Roots of Democratic Stability." In *Democracy in Developing Countries: Latin America* (Volume Four), ed. Larry Diamond et al. Boulder: Lynne Reiner.

Borges, Fabián. 2004. "Low Season Starts in High Spirits after Banner Year." *Tico Times,* May 21, 2004: S–2.

Boza, Mario. 1981. *El Sistema de Parques Nacionales de Costa Rica: Una Década de Desarollo.* San José, C.R.: Editorial UNED.

—— 1993. "Conservation in Action: Past, Present, and Future of the National Park System of Costa Rica." *Conservation Biology* 7(2): 239–47.

Boza, Mario, et al. 1995. "Costa Rica is a Laboratory, Not Ecotopia." *Conservation Biology* 9(3): 684–5.

Bravo, Hernán. 1992. "La Utilización Racional de los Recursos Naturales y el Desarollo de Costa Rica." In *Desarrollo Sostenible y Políticas Económicas en América Latina,* Omar Segura. San José, C.R.: DEI.

Brockington, Dan. 2002. *Fortress Conservation: The Preservation of the Mkomazi Game Preserve, Tanzania.* Bloomington: Indiana University Press.

Brosius, J. Peter. 1999. "Analyses and Interventions: Anthropological Engagements with Environmentalism." *Current Anthropology* 40(3): 277–309.

Budiansky, Stephen. 1995. *Nature's Keepers: The New Science of Nature Management.* New York: Free Press.

Budowski, Gerardo. 1985. *La Conservación Como Instrumento Para el Desarrollo.* San José, Costa Rica: Editorial UNED.

Budoswki, Tamara. 1992. "Ecotourism Costa Rican Style." In *Toward a Green Central America: Integrating Conservation and Development,* eds. Valerie Barzetti and Yvette Rovinski. W. Hartford, Conn.: Kumarian Press.

Bulgarelli, V. 1976. Letter to J. Tosi, CCT. Procaduría de la República, No. 33—PA 76. Archivos Nacionales de Costa Rica. Folder No. 474, Servicio de Parques Nacionales.

Burlingame, Leslie. 2000. "Conservation in the Monteverde Zone: Contributions of Conservation Organizations." In *Monteverde: Ecology and Conservation of a Tropical Cloud Forest*, eds. Nalini Nadkarni and Nat Wheelwright. Oxford: Oxford University Press.

Burnett, John. 1997. "Ecotourism." *All Things Considered*. National Public Radio. 3 September 1997.

Calderón, Rafael Angel. 1990. "Proclama Sobre el Ambiente Hacia un Nuevo Ordern Ecológico de Cooperación Internacional." Speech Presented by the President of the Republic, 14 December 1990. San José, Costa Rica.

Campbell, Lisa. 2002. "Conservation Narratives in Costa Rica: Conflict and Co-existence." *Development and Change*, Volume 33, no. 1 (2002): 29–56.

Carranza, Carlos, et al. 1996. *Valoración de los Servicios Ambientales de los Bosques de Costa Rica*. San José, C.R.: Centro Científico Tropical.

Carroll, C.R. 1992. "Ecological Management of Sensitive Areas." In *Conservation Biology: The Theory and Practice of Nature Conservation, Preservation and Management*, eds. Peggy Fiedler and Subodh Jain. New York: Chapman and Hall.

Carroll, Thomas F. 1992. *Intermediary NGOs: The Supporting Link in Grassroots Development*. West Hartford, CT: Kumarian.

Castro, Raúl. 1996. "Decentralización y Municipalización de la gestión ambiental." *Ambientico* 38: 2–7.

Castro, Sylvia and F. Willink. 1989. *San Ramón: Economía y Sociedad 1900–1948*. San Ramón, C.R.: Universidad de Costa Rica, Sede de Occidente.

Cater, Erlet. 1994. "Introduction." In *Ecotourism: A Sustainable Option?* Cater, E. and Gwen Lowman. Chichester, UK: John Wiley & Sons.

Caufield, Catherine. 1984/91. *In the Rainforest: Report from a Strange, Beautiful, Imperiled World*. Chicago: University of Chicago Press.

CCT (Centro Científico Tropical). 1968. "Estudio sobre capacidad de uso de la tierra en la región norte de Alajuela y Heredia." San José, Costa Rica.

—— 1986. "Saving the Cloud Forests: The Campaign for the Peñas Blancas Watershed." World Wildlife Fund File 6080, Washington, DC.

—— 1995. "Monteverde-Gulf of Nicoya Biological Corridor Costa Rica." San José, Costa Rica: Tropical Science Center.

Chamberlain, Francisco. 1993. "The Monteverde Cloud Forest Preserve Chapter." Leadership for Environment and Development Program. First International Session, Second Cohort (1993–1995). The Costa Rica Case Study. San José, Costa Rica. June 10–21, 1993.

—— 2000. "Pros and Cons of Ecotourism." In *Monteverde: Ecology and Conservation of a Tropical Cloud Forest,* eds. Nalini Nadkarni and Nat Wheelwright. Oxford: Oxford University Press.

Chambers, Robert. 1993. *Challenging the Professions: Frontiers for Rural Development.* London: Intermediate Technology Publications.

Chase, Alton. 1987. *Playing God in Yellowstone: The Destruction of America's First National Park.* San Diego: Harcourt Brace Jovanovich.

Children's Rainforest U.S. 1994. Informational Brochure. Lewiston, Maine.

Clifford, James. 2001. "Indigenous Articulations." *The Contemporary Pacific* 13(2): 468–90.

Colchester, Marcus. 1993. "Colonizing the Rain Forests: The Agents and Causes of Deforestation." In *The Struggle for Land and the Fate of the Forests,* eds. Marcus Colchester and Larry Lohmann. N.J.: Zed Books.

Collard, Sneed. 1997. *Monteverde: Science and Scientists in a Costa Rican Cloud Forest.* New York: Franklin Watts.

Collinson, Helen, ed. 1996. *Green Guerrillas: Environmental Conflicts and Initiatives in Latin America and the Caribbean.* London: Latin American Bureau.

Craik, Jennifer. 1997. "The Culture of Tourism." In *Touring Cultures: Transformations of Travel and Theory,* eds. Chris Rojek and John Urry. London: Routledge.

Crick, Malcolm. 1989. "Representations of International Tourism in the Social Sciences: Sun, Sex, Sights, Savings and Servility." *Annual Review of Anthropology* 18: 307–44.

Croll, Elisabeth and David Parkin. 1992. "Anthropology, the Environment and Development." In *Bush Base: Forest Farm: Culture, Environment, and Development,* eds. Elisabeth Croll and David Parkin. London: Routledge.

Cronon, William, ed. 1996. *Uncommon Ground: Rethinking the Human Place in Nature.* New York: Norton.

CTASE (Colegio Técnico Agropecuario de Santa Elena). 1982. "Proyecto Parcelación, Finca Bee, S.A." Colegio Técnico Agropecuario de Santa Elena, Monte Verde, Costa Rica.

Cummings, T. 1989. "An Inventory of Formerly Settled Lands in the Peñas Blancas Valley. Monteverde Cloud Forest Reserve, Costa Rica. Implications for Reserve Management." TRI Working Paper, Part Three of Three. Yale School of Forestry and Environmental Studies. New Haven, Conn.

Dahl, Robert. 1999. "Can International Organizations be Democratic? A Skeptic's View." In *Democracy's Edges* edited by Ian Shapiro and Cadiano Hacker-Cordón. Cambridge: Cambridge University Press.

Davis, Susan. 1996. *Spectacular Nature: Corporate Culture and the Sea World Experience.* Berkeley: University of California Press.

DGF (Dirección General Forestal). 1987. "Reserva Forestal Arenal: Presente o Futuro?" Ministerio de Agricultura y Ganadería, Direccion General Forestal. San José, Costa Rica, November 1987.

Dove, Michael. 1993. "A Revisionist View of Tropical Deforestation and Development." *Environmental Conservation* 20(1): 17–24, 56.

Dowie, Mark. 1995. *Losing Ground: American Environmentalism at the Close of the Twentieth Century.* Cambridge: MIT Press.

Dwyer, V. 1988. "Cheap Conservation at $25 an Acre." *MacLean's* Vol. 101: 52.

Echeverría, Jaime, et al. 1995. "Valuation of non-priced amenities provided by the biological resources within the Monteverde Cloud Forest Preserve, Costa Rica." *Ecological Economics* 13: 43–52.

Edelman, Marc. 1992. *The Logic of the Latifundio: The Large Estates of Northwestern Costa Rica since the Late Nineteenth Century.* Stanford: Stanford University Press.

—— 1995. "Rethinking the Hamburger Thesis: Deforestation and the Crisis of Central America's Beef Exports." In *The Social Causes of Environmental Destruction in Latin America,* eds. Michael Painter and William Durham. Ann Arbor: University of Michigan Press.

—— 1999. *Peasants Against Globalization: Rural Social Movements in Costa Rica.* Stanford: Stanford University Press.

Edelman, Marc and Joanne Kenen, eds. 1989. *The Costa Rica Reader.* New York: Grove Weidenfeld.

Edwards, Michael and David Hulme. 1996. "Too Close for Comfort? The Impact of Official Aid on Nongovernmental Organizations." *World Development* 24(6): 961–73.

Ellen, Roy. 1986. "What Black Elk left unsaid: On the Illusory Images of Green Primitivism." *Anthropology Today* 2(6): 8–12.

Ellis, Jeffrey. 1996. "On the Search for a Root Cause: Essentialist Tendencies in Environmental Discourse." In *Uncommon Ground: Rethinking the Human Place in Nature,* ed. William Cronon. New York: Norton.

Escobar, Arturo. 1995. *Encountering Development: The Making and Unmaking of the Third World.* Princeton: Princeton University Press.

—— 1997. "Cultural Politics and Biological Diversity: State, Capital, and Social Movements in the Pacific Coast of Colombia." In *Between Resistance and Revolution: Cultural Politics and Social Protest,* eds. Richard Fox and Orin Starn. New Brunswick, NJ: Rutgers University Press.

Escobar, Francisco. 1977. *Sociedad y Comunidad Rural: Una Perspectiva de la Comunidad Rural Costarricense.* San José, C.R.: Ministerio de Cultura y Deportes.

Escofet, Guillermo. 1998. "Forest Report Disputed." *Tico Times,* 20 March 1998.

Esteva, Gustavo and Madhu Suri Prakash. 1998. *Grassroots Postmodernism: Remaking the Soil of Cultures.* London: Zed Books.

Evans, Sterling. 1999. *The Green Republic: A Conservation History of Costa Rica.* Austin: University of Texas Press.

Fairhead, James and Melissa Leach. 1996. *Misreading the African Landscape: Society and Ecology in a Forest-Savanna Mosaic.* Cambridge: Cambridge University Press.

Fallas Baldi, O. 1992. *Modelos de Desarollo y Crisis Ambiental en Costa Rica.* Serie Cuadernos de Estudio. San José, Costa Rica: Asociación Ecologista Costarricense.

Fennel, David and Paul Eagles. 1990. "Ecotourism in Costa Rica: A Conceptual Framework." *Journal of Parks and Recreation Administration* 8(1), Spring 1990, pp. 23–34.

Ferguson, James. 1994. *The Anti-Politics Machine: 'Development,' Depoliticization and Bureaucratic Power in Lesotho.* Minneapolis: University of Minnesota Press.

Fernández, Mario. 1989. "Acceso a la Tierra y Reproducción de Campesinado en Costa Rica." *Ciencias Sociales* 43: 31–41.

Fiedler, Peggy and Subodh Jain, eds. 1992. *Conservation Biology: The Theory and Practice of Nature Conservation, Preservation and Management.* New York: Chapman and Hall.

Figueroa, L.F. 2002. *Modelo de Planeación Estratégica Financiera para la Reserva Biológica Bosque Nuboso de Monteverde, Años 2002–2006.* Master's Thesis in Business Administration, University of Costa Rica, System of Post-Graduate Studies. San José, Costa Rica.

Fisher, Julie. 1993. *The Road from Rio: Sustainable Development and the Nongovernmental Movement in the Third World.* Westport, Conn: Praeger.

Fisher, William. 1997. "Doing Good? The Politics and Antipolitics of NGO Practices." *Annual Review of Anthropology* 26: 249–64.

Fisher, William, ed. 1995. *Toward Sustainable Development? Struggling Over India's Narmada River.* Armonk, NY: ME Sharpe.

Forsyth, Adrian. 1988. "The Lessons of Monteverde." *Equinox* March/April 1988, pp. 56–61.

Fournier, Luis. 1985. *Ecología y Desarrollo en Costa Rica.* San José, Costa Rica: EUNED.

———— 1991. *Desarrollo y Perspectiva del Movimiento Conservacionista Costarricense.* San José, Costa Rica: Editorial UCR.

———— 1993. *Recursos Naturales.* San José, C.R.: Editorial UNED.

Fox, Richard and Orin Starn. 1997. "Introduction." In *Between Resistance and Revolution: Cultural Politics and Social Protest,* ed. Richard Fox and Orin Starn. New Brunswick, NJ: Rutgers University Press.

Fundación Centro Ecológico Bosque Nuboso de Monteverde. n.d. "Fundación Centro Ecológico del Bosque Nuboso de Monteverde." Santa Elena de Monte Verde, Costa Rica.

———— 1992. Legal Constitution. Santa Elena de Monte Verde, Costa Rica.

Fundación Neotrópica. 1988. *Desarrollo Socioeconómico y el Ambiente Natural de Costa Rica: Situación Actual y Perspectivas.* San José, C.R.: Fundación Neotrópica.

Fürst, Edgar and Wolfgang Hein. 2002. *Turismo de Larga Distancia y Desarollo Regional en Costa Rica.* San José, Costa Rica: DEI.

Gámez, Rodrigo. 1991. "Biodiversity Conservation Through Facilitation of its Sustainable Use: Costa Rica's National Biodiversity Institute." *Tree* 6, no. 12 (December 1991): 377–78.

———— 1999. *De Biodiversidad, Gentes y Utopías: Reflexiones en los 10 Años del INBio.* Heredia, Costa Rica: Instituto Nacional de Biodiversidad.

Gámez, Rodrigo and Alvaro Ugalde. 1988. "Costa Rica's National Park System and the Preservation of Biodiversity: Linking Conservation with Socio-Economic Development." In *Tropical Rainforests: Diversity and Conservation*, eds. Frank Almeda and Catherine M. Pringle. San Francisco: California Academy of Sciences.

García, E. 2002. "Bosques de Costa Rica: Mucho Más que Arboles." In *El Contexto Ecológico de Costa Rica a Finales del Siglo XX,* Sonia Amador. San José, Costa Rica: Editorial de la Universidad de Costa Rica.

García, Randall. 2004. "Falta Armonizar Turismo y Conservación." *Ambientico* 126, Marzo 2004: 4–5.

Global Volunteers. 1996. "Global Volunteers: Adventures in Service." Promotional brochure. Global Volunteers: St. Paul, MN.

Gómez, Luis Diego. 2000. "Foreword." In *Monteverde: Ecology and Conservation of a Tropical Cloud Forest,* eds. Nalini Nadkarni and Nat Wheelwright. Oxford: Oxford University Press.

Gómez, Luis Diego and J.M. Savage. 1983. "Searchers on that Rich Coast: Costa Rican Field Biology 1400–1980." In *Costa Rican Natural History,* ed. Daniel Janzen. Chicago: University of Chicago.

Gómez-Pompa, Arturo and Andrea Kaus. 1992. "Taming the Wilderness Myth." *BioScience* 42(4): 271–9.

Gottlieb, Robert. 1993. *Forcing the Spring: The Transformation of the American Environmental Movement.* Washington, DC: Island Press.

Green, G. and J. Barborak. 1987. "Conservation for Development: Success Stories from CentralAmerica." *Commonwealth Forestry Review* 66: 91–102.

Griffith, Katherine, et al. 2000. "Agriculture in Monteverde: Moving Toward Sustainability." In *Monteverde: Ecology and Conservation of a Tropical Cloud Forest,* eds. Nalini Nadkarni and Nat Wheelwright. Oxford: Oxford University Press.

Grosby, Samantha. 2000. "The Changing Face of Tourism." In *Monteverde: Ecology and Conservation of a Tropical Cloud Forest,* eds. Nalini Nadkarni and Nat Wheelwright. Oxford: Oxford University Press.

Gudeman, Steven and Alberto Rivera. 1990. *Conversations in Colombia: The Domestic Economy in Life and Text.* Cambridge: Cambridge University Press.

Gudmunson, Lowell. 1983. *Hacendados, Políticos y Precaristas: La Ganadería y el Latifundismo Guanacasteco, 1800–1950.* San José, C.R.: Editorial Costa Rica.

Guevara, J.D. 1990. "Pequeños Aliados de Monteverde." *La Nación* 2/2/90. Viva p. 2.

Guha, Ramachandra. 1997a. "The Authoritarian Biologist and the Arrogance of Anti-Humanism: Wildlife Conservation in the Third World." *The Ecologist* 27(1): 14–20.

—— 1997b. "The Environmentalism of the Poor." *In Between Resistance and Revolution: Cultural Politics and Social Protest,* ed. Richard Fox and Orin Starn. New Brunswick, NJ: Rutgers University Press.

—— 2000. *Environmentalism: A Global History.* New York: Longman.

Guha, Ramachandra and Joan Martinez-Alier. 1997. *Varieties of Environmentalism: Essays North and South.* London: Earthscan.

Guindon, Carlos. 1996. "The Importance of Forest Fragments to the Maintenance of Regional Biodiversity in Costa Rica." In *Forest Patches in Tropical Landscapes,* eds. John Shelhas and R. Greenberg. Washington, DC: Island Press.

—— 2000. "The Importance of Pacific Slope Forest for Maintaining Regional Biodiversity." In *Monteverde: Ecology and Conservation of a Tropical Cloud Forest,* eds. Nalini Nadkarni and Nat Wheelwright. Oxford: Oxford University Press.

Gutierrez, Wilfredo. 1983. "Caracterización de los Sistemas Predominantes con Enfasis en el Componente Bovino, en Fincas Familiares de Cariari y Monteverde, Costa Rica." M.S. Thesis, Department of Animal Production. Turrialba, C.R.: CATIE.

Hagen, Joel. 1985. *An Entangled Bank: The Origins of Ecosystem Ecology.* New Brunswick, NJ: Rutgers University Press.

Hall, Carolyn. 1985. *Costa Rica: A Geographical Interpretation in Historical Perspective.* Dellplain Latin American Studies, No. 17. Boulder: Westview Press.

Hannerz, Ulf. 1990. "Cosmopolitans and Locals in World Culture." *Theory, Culture & Society* 7: 237–51.

Hansen, Anders, ed. 1993. *The Mass Media and Environmental Issues.* Leicester: Leicester University Press.

Harper, J. 1992. "Foreword." In *Conservation Biology: The Theory and Practice of Nature Conservation, Preservation and Management.* New York: Chapman and Hall.

Hartshorn, Gary, et al. 1982. *Costa Rica: A Country Environmental Profile, A Field Study.* San José, C.R.: Tropical Science Center, USAID.

Harvey, Celia. 2000. "Windbreaks as Habitats for Trees." In *Monteverde: Ecology and Conservation of a Tropical Cloud Forest*, eds. Nalini Nadkarni and Nat Wheelwright. Oxford: Oxford University Press.

Head, Suzanne and Robert Heinzman. 1990. *Lessons of the Rainforest.* San Francisco: Sierra Club Books.

Hecht, Susan and Alexander Cockburn. 1990. *The Fate of the Forest: Developers, Destroyers and Defenders of the Amazon.* New York: Harper Perennial.

Herlihy, Peter. 1992. "'Wildlands' Conservation in Central America during the 1980s: A Geographical Perspective." *Conference of Latin American Geographers* 17/18, p. 31–43.

Hilje, Luko, et al. 1987. *El Uso de Plaguicidas en Costa Rica.* San José, Costa Rica. Editorial UNED.

Hilje, Luko, et al. 2002. *Los Viejos y Los Arboles.* Heredia, Costa Rica: Instituto Nacional de Biodiversidad and Editorial de la Universidad de Costa Rica.

Hilje Q., B. 1987. *Colonización Agrícola de Tilarán, 1880–1950.* Heredia, C.R.: Universidad Nacional Autónoma. Tésis de Licenciatura en Historia.

Hill, Carol. 1990. "The Paradox of Tourism in Costa Rica." *Cultural Survival Quarterly* 14(1): 14–19.

Hollinshead, Keith. 1999. "Tourism as Public Culture: Horne's Ideological Commentary on the Legerdemain of Tourism." *International Journal of Tourism Research* 1: 267–92.

Honey, Martha. 1994a. *Hostile Acts: U.S. Policy in Costa Rica in the 1980s.* Gainesville: University Press of Florida.

—— 1994b. "Paying the Price of Ecotourism: Two Pioneer Biological Reserves Face the Challenges Brought by a Recent Boom in Tourism." *Américas* 46(6): 40–7.

—— 1999. *Ecotourism and Sustainable Development: Who Owns Paradise?* Washington, DC: Island Press.

Horne, Donald. 1992. *The Intelligent Tourist.* McMahon's Point, Australia: Margaret Gee Holdings.

Hotel Heliconia Monteverde. n.d. "Hotel Heliconia Monteverde, Costa Rica." Cerro Plano de Monteverde, Costa Rica.

Howard, Marian P. 1989. *An Alternative Way of Being: The Ethnographic Study of a Quaker Community in the Cloud Forest of Costa Rica—1987.* Ph.D. Dissertation, Teachers College, Columbia University.

Hunter, J. Robert. 1994. "Is Costa Rica Truly Conservation-Minded?" *Conservation Biology* 8(2): 592–5.

ICE (Instituto Costarricense de Electricidad). 1978. "Reserva Forestal Arenal: Reseña Histórica." Oficina de Estudios Especiales. Dirección de Electrificación. ICE. San José, Costa Rica.

ICT (Instituto Costarricense de Turismo). 1994. "Santa Elena y Monteverde: Naturaleza y mucho más." *Disfrutemos Costa Rica.* Año 3, pp. 3–7. September 1994. San José, Costa Rica.

—— 2004. "Regiones y Estimación de Visitación, 2003–4." Statistical Section, San José, Costa Rica.

Igoe, James. 2004. *Conservation and Globalization: A Study of National Parks and Indigenous Communities from East Africa to South Dakota.* Belmont, CA: Thomson/Wadsworth.

Ingold, Timothy. 1993. "Globes and Spheres: The Topology of Environmentalism." In *Environmentalism: The View from Anthropology,* ed. Kay Milton. New York: Routledge.

IUCN (International Union for the Conservation of Nature). 1980. *The World Conservation Strategy.* Geneva: IUCN, UN Environment Program, WWF.

Janzen, Daniel. 1986. "The Future of Tropical Ecology." *Annual Review of Ecology and Systematics* 17: 305–24.

—— 1991. "How to Save Tropical Biodiversity: The National Institute of Biodiversity of Costa Rica." *American Entomologist* Fall 1991: 159–171.

Johnston, Barbara Rose. 1990. "Introduction: Breaking out of the Tourist Trap." *Cultural Survival Quarterly* 14(1), 2–5.

Jones, Jeffrey. 1990. *Colonization and Environment: Land Settlement Projects in Central America.* Tokyo: United Nations University Press.

Jukofsky, Diane. 1993. "Mystical Messenger." *Nature Conservancy.* November/December 1993.

Kabeer, Naila. 1994. *Reversed Realities: Gender Hierarchies in Development Thought.* London: Verso.

Kamuaro, Ole. 1996. "Ecotourism: Suicide or Development?" *Voices from Africa #6 Sustainable Development.* United Nations Government Liaison Service.

Kearney, Michael. 1996. *Reconceptualizing the Peasantry: Anthropology in Global Perspective.* Boulder: Westview Press.

Kinsman, Sharon. 1991. "Education and Empowerment: Conservation Lessons from Children." *Conservation Biology* 5(1): 9–10.

Kirshenblatt-Gimblett, Barbara. 1998. *Destination Culture: Tourism, Museums, and Heritage.* Berkeley: University of California Press.

Korten, Alicia. 1997. *Ajuste Estructural en Costa Rica: Una Medicina Amarga.* San José, Costa Rica: DEI/Food First.

Korten, David. 1990. *Getting to the 21st Century: Voluntary Action and the Global Agenda.* West Hartford, Conn.: Kumarian Press.

Kuzmier, K. 1992. *A Development Strategy and Management Plan for a Community-Run Ecotourism Project in Santa Elena, Costa Rica.* M.A. Thesis, School of the Environment, Duke University, Durham, North Carolina.

Laarman, Jan and Richard Perdue. 1989. "Science Tourism in Costa Rica." *Annals of Tourism Research* 16: 205–15.

Leach, Melissa and Robin Mearns. 1996. "Environmental Change and Policy: Challenging Received Wisdom in Africa." In *The Lie of the Land: Challenging the Received Wisdom on the African Environment,* eds. Melissa Leach and Robin Mearns. Oxford: The International African Institute.

Lefebvre, Henri. 1991. *The Production of Space.* Cambridge, MA: Blackwell.

Leitinger, Ilse A. 1997. "Long-Term Survival of a Costa Rican Women's Crafts Cooperative: Approaches to Problems of Rapid Growth at CASEM in the Santa Elena-Monteverde Region." In *The Costa Rican's Women's Movement: A Reader,* ed. Ilse A. Leitinger. Pittsburgh: University of Pittsburgh Press.

Lindberg, Kreg and Donald Hawkins. 1993. *Ecotourism: A Guide for Planners and Managers.* Bennington, VT: The Ecotourism Society.

Lomborg, Bjorn. 2001. *The Skeptical Environmentalist: Measuring the Real State of the World.* Cambridge: Cambridge University Press.

Lummis, C. Douglas. 1996. *Radical Democracy.* Ithaca: Cornell University Press.

MacArthur, R.H. and E.O. Wilson. 1967. *The Theory of Island Biogeography.* Princeton: Princeton University Press.

MacCannell, Dean. 1989. *Empty Meeting Grounds: The Tourist Papers.* New York: Routledge.

Macdonald, Laurie. 1995. "NGOs and the Problematic Discourse of Participation: Cases from Costa Rica." In *Debating Development Discourse: Institutional and Popular Perspectives,* eds. David Moore and Gerald Schmitz. London: Macmillan Press.

—— 1997. *Supporting Civil Society: The Political Role of Non-Governmental Organizations in Central America.* London: Macmillan Press.

MAG (Ministerio de Agricultura y Ganadería). 1985. "Proyecto Silvopastoral Colegio Técnico Agropecuario Sta. Elena, Puntarenas." Dirección General Forestal, Departamento Repoblación Forestal. San José, Costa Rica.

—— 1996. Unpublished statistical information on Monte Verde farms. MAG satellite office, Santa Elena, Costa Rica.

Marozzi, M. 2002. "Los Límites del Modelo Inbio." *Ambientico* 100: 10–11.

Martin, Emily. 1994. *Flexible Bodies: Tracking Immunity in American Culture from the Days of Polio to the Age of AIDS.* Boston: Beacon Press.

Martinez-Alier, Joan. 1990. "Ecology and the Poor: A Neglected Dimension of Latin American History." *Journal of Latin American Studies* 23: 621–39.

McCoy, E. and K. Shrader-Frechette. 1992. "Community Ecology, Scale and the Instability of the Stability Concept." In *Philosophy of Science Association*

*1992,* eds. Forbes and Hull. E. Lansing, Michigan: Philosophy of Science Association 1992. Vol. 1: 84–99.

McDade, Lucinda and Gary Hartshorn. 1994. "La Selva Biological Station." In *La Selva: Ecology and Natural History of a Neotropical Rainforest,* ed. Lucinda McDade, et al. Chicago: University of Chicago Press.

McIntosh, Robert P. 1985. *The Background of Ecology: Concept and Theory.* Cambridge: Cambridge University Press.

McKibben, Bill. 1989. *The End of Nature.* New York: Random House.

MCFP (Monteverde Cloud Forest Preserve). 1993. "Proposal for Environmental Education Program." Unpublished document.

MCL (Monteverde Conservation League). 1986. "Saving the Monteverde Cloud Forest." Proposal to the World Wildlife Fund. WWF Files, Washington, DC.

—— 1986. *Tapir Tracks: A Newsletter of the Monteverde Conservation League.* Vol 1(2). Nov./Dec. 1986.

—— 1988. *Tapir Tracks: A Newsletter of the Monteverde Conservation League.* Vol. 3(1). August 1988.

—— 1991. "Barnens Regnskog, Suecia." *Tapir Tracks: A Newsletter of the Monteverde Conservation League.* Vol. 6(2), Mayo-Julio 1991.

—— 1992a. "Editorial." *Tapir Tracks: A Publication of the Monteverde Conservation League* 7(1): 3–8. Monteverde Conservation League, Monteverde, Costa Rica.

—— 1992b. "A Conservation and Sustainable Development Project for Research in the International Children's Rain Forest at San Gerardo Arriba, Costa Rica." May 1992. MCL archives—San Gerardo folder.

—— 1994a. "Informe Anual 1994." Asamblea General, 11 de noviembre 1994.

—— 1994b. "Report to the Board of Directors Alex C. Walker Educational and Charitable Foundation." MCL archives—San Gerardo folder.

—— n.d. "The Monteverde Conservation League—Fact Sheet." Monteverde Conservation League.

McLaren, Deborah. 2003. *Rethinking Tourism and Ecotravel,* 2nd Edition. West Hartford, CT: Kumarian Press.

McNaughton, S.J. 1989. "Ecosystems and Conservation in the Twenty-First Century." In *Conservation for the Twenty-First Century,* ed. David Western and M. Pearl. Oxford: Oxford University Press.

Mendenhall, Mary. 1995. *Monteverde.* Canadian Quaker Pamphlet Series No. 42. Argenta, B.C., Canada: Argenta Friends Press.

Milton, Kay, ed. 1993. *Environmentalism: The View from Anthropology.* ASA Monographs 32. London: Routledge.

Milton, Kay. 1996. *Environmentalism and Cultural Theory.* London: Routledge.

Ministerio de Economía, Industria y Comercio. 1974. *Censo Agropecuario 1973.* San José, C.R.: Dirección General de Estadísticas y Censos.

—— 1987. *Censo Agropecuario 1984.* San José, C.R.: Dirección General de Estadísticas y Censos.

—— 1994. *Costa Rica: Cálculo de Población por Provincia, Cantón y Distrito.* San José, C.R.: Dirección de Estadísticas y Censos.

Montero, Victor and Zeidy Zarate. 1991. *Comunidades Sociobióticas: un modelo operativo de sostenibilidad.* Thesis for Licenciatura in Architecture. University of Costa Rica, School of Architecture.

Mora Castellanos, Eduardo. 1993. *Claves del Discurso Ambientalista.* Heredia, Costa Rica: Edtorial UNA.

—— 1998. *Naturaleza, QuéHerida Mía.* Heredia, Costa Rica: Ambientico Ediciones.

Morell, Virginia. 1999. "The Sixth Extinction." *National Geographic* Feb. 1999, pp. 42–56.

Morgan, Lynn. 1993. *Community Participation in Health: The Politics of Primary Care in Costa Rica.* Cambridge: Cambridge University Press.

Morrison, Polly J. 1994. "The Monteverde Area of Costa Rica: A Case Study of Ecotourism Development." M.A. Thesis, University of Texas. Latin American Studies and Community Development. Austin, Texas.

Munt, Ian. 1994. "Eco-Tourism or ego-tourism?" *Race & Class* 36(1): 49–60.

Murphy, P.E. 1985. *Tourism: A Community Approach.* New York: Methuen.

Musinsky, J. 1991. "The Design of Conservation Corridors in Monteverde, Costa Rica." TRI Working Paper #60. Yale University School of Forestry and Environmental Studies and Tropical Resources Institute. December 1991.

Myers, Norman. 1984/92 *The Primary Source: Tropical Forests and our Future.* New York: Norton.

Nadkarni, Nalini. 2000. "Scope of Past Work." In *Monteverde: Ecology and Conservation of a Tropical Cloud Forest,* eds. Nalini Nadkarni and Nat Wheelwright. Oxford: Oxford University Press.

Nadkarni, Nalini and Nat Wheelwright, eds. 2000. *Monteverde: Ecology and Conservation of a Tropical Cloud Forest.* Oxford: Oxford University Press.

Nations, James and Daniel Komer. 1987. "Rainforests and the Hamburger Society." *The Ecologist* 17(4/5): 161–7.

Neumann, Roderick. 1998. *Imposing Wilderness: Struggles over Livelihood and Nature Preservation in Africa.* Berkeley: University of California Press.

Nielsen, Karen and Deborah DeRosier. 2000. "Windbreaks as Corridors for Birds." In *Monteverde: Ecology and Conservation of a Tropical Cloud Forest,* eds. Nalini Nadkarni and Nat Wheelwright. Oxford: Oxford University Press.

Nygren, Anja. 1993. *El Bosque y la Naturaleza en la Percepción del Campesino Costarricense: Un Estudio de Caso.* Turrialba, Costa Rica: CATIE, Programa Manejo de Recursos Naturales.

—— 1995. "Deforestation in Costa Rica: An Examination of Social and Historical Factors." *Forest and Conservation History* 39: 27–35.

Obando, Vilma. 2002. *Biodiversidad en Costa Rica: Estado del Conocimiento y Gestión.* Heredia, Costa Rica: Instituto Nacional de Biodiversidad.

O'Connor, Geoffrey. 1997. *Amazon Journal: Dispatches from a Vanishing Frontier.* New York: Dutton.

Orlove, Benjamin and Stephen Brush. 1996. "Anthropology and the Conservation of Biodiversity." *Annual Review of Anthropology* 25: 329–52.

Owen, D.F. 1974. *What is Ecology?* Oxford: Oxford University Press.

Painter, M. and W. Durham. 1995. *The Social Causes of Environmental Destruction in Latin America.* Ann Arbor: University of Michigan Press.

Park, Christopher. 1992. *Tropical Rainforests.* London: Routledge.

Parkin, David. 1996. "Introduction: The Power of the Bizarre." In *The Politics of Cultural Performance,* eds. David Parkin, Lionel Caplan and Humphrey Fisher. Providence: Berghahn Books.

Patent, Dorothy Hinshaw. 1996. *Children Save the Rain Forest.* New York: Cobblehill Books/Dutton.

Paulson, Susan and Lisa Gezon. 2004. *Political Ecology Across Space, Scales and Social Groups.* New Brunswick: Rutgers University Press.

Pérez Sainz, J.P. 1999. *Mejor Cercanos que Lejanos: Globalización, Autogeneración de Empleo y Territorialidad en Centroamerica.* San José, Costa Rica: FLACSO.

Perlman, Dan and Glenn Adelson. 1997. *Biodiversity: Exploring Values and Priorities in Conservation.* Oxford: Blackwell Science.

Peters, Pauline. 1996. "Who's Local Here?: The Politics of Participation in Development." *Cultural Survival Quarterly* 20(3): 22–5.

Peters, Robert H. 1991. *A Critique for Ecology.* Cambridge: Cambridge University Press.

Pickett, Stewart et al. 1992. "A New Paradigm in Ecology: Implications for Conservation Biology Above the Species Level." In *Conservation Biology: The Theory and Practice of Nature Conservation, Preservation and Management,* eds. Peggy Fiedler and Subodh Jain. New York: Chapman and Hall.

Pickett, Stewart et al. 1994. *Ecological Understanding: The Nature of Theory and the Theory of Nature.* San Diego: Academic Press.

Pimm, Stuart. 1991. *Balance of Nature? Ecological Issues in the Conservation of Species and Communities.* Chicago: University of Chicago Press.

Piot, Charles. 1999. *Remotely Global: Village Modernity in West Africa.* Chicago: University of Chicago Press.

Place, Susan, ed. 1993. *Tropical Rainforests: Latin American Nature and Society in Transition.* Wilimington, DE: Scholarly Resources.

Porras, Anabelle and Beatriz Villareal. 1993. *Deforestación en Costa Rica: Implicaciones Sociales, Económicas, y Legales.* San José, C.R.: Imprenta Nacional.

Pounds, Alan and Marty Crump. 1994. "Amphibian Declines and Climate Disturbance: The Case of the Golden Toad and the Harlequin Frog." *Conservation Biology* 8(1): 72–85.

Pounds, Alan, et al. 1999. "Biological Response to Climate Change on a Tropical Mountain." *Nature* 398: 611–15.

Powell, George. 1974. "Monteverde Project." Application to the World Wildlife Fund. From files of World Wildlife Fund #6080, U.S. Washington, DC.

—— 1989. "To Save a Forest: The Monteverde Preserve's First Year." In *Tapir Tracks: The Newsletter of the Monteverde Conservation League.* July/October 1989.

Powell, George and Robin Bjork. 1993. "Altitudinal Migrations and Habitat Linkages in Montane Environments of Costa Rica: A Call for the Design of Altitudinal Corridors." In *Corredores Conservacionistas en la Región Centroamericana,* ed. Vega. Gainesville, FL: Tropical Research and Development, Inc.

Quesada Mateo, Carlos. 1992. *El Canje de la Deuda Externa para Promover la Conservación de los Recursos Naturales.* Rome: FAO.

Quesada Mateo, Carlos and Vivienne Solís, eds. 1988. *Memoria 1er Congreso Estrategia de Conservación para el Desarrollo Sostenible de Costa Rica.* San José, Costa Rica: MIRENEM, IUCN, WWF.

Rachowiecki, Robert. 2002. *Costa Rica: A Travel Survival Kit.* 5th edition. Berkeley: Lonely Planet.

Radulovich, Ricardo. 1988. "Degradación Ambiental en Costa Rica." *Agronomía Costarricense* 12(2): 253–71.

Rahnema, Majid. 1990. "Participatory Action Research: The 'Last Temptation of Saint' Development." *Alternatives* XV (1990): 199–226.

—— 1992. "Participation." In *The Development Dictionary: A Guide to Knowledge and Power,* ed. Wolgang Sachs. London: Zed Books.

Ramírez Flores, E. 2004. "Monteverde Clama por Vía Asfaltada." *Semanario Universidad* 20 de Mayo 2004, p. 9.

Raven, Peter. 1986. "The Urgency of Tropical Conservation." *The Nature Conservancy News* January-March 1986.

Reid, Walter and Kenton Miller. 1989. *Keeping Options Alive: The Scientific Basis of Conserving Biodiversity.* World Resources Institute.

Roberts, J. Timmons and Nikki Demetria Thanos. 2003. *Trouble in Paradise: Globalization and Environmental Crises in Latin America.* New York: Routledge.

Robinson, B. 1988. "The Costa Rican Connection." Equinox March/April 1988, pp. 8–10.

Rodríguez, Carlos. 1993. *Tierra de Labriegos: Los Campesinos en Costa Rica desde 1950.* San José, C.R.: FLACSO.

Rodriguez, Silvia. 1992. "Papel de la etica en la patentización de la biodiversidad." *Praxis* 43–4. Departamento de Filosofía Universidad Nacional Autonoma, Heredia, Costa Rica.

——— 1994. *Conservación, Contradicción y Erosion de Soberanía: El Estado Costarricense y las Areas Naturales Protegidas (1970–1992).* PhD Thesis, University of Wisconsin Madison, Spanish Translation.

Roe, Emery. 1991. "Development Narratives, Or Making the Best of Blueprint Development." *World Development* 19(4): 287–300.

Rojas, C. 1992. "Monteverde: estudio inicial de los nexos entre la reserva, el turismo y la comunidad local." Unpublished Manuscript. RARE Center for Tropical Conservation and Centro Científico Tropical, San José, Costa Rica.

Rojas Bolaños, Manuel. 1990. "La Democracia Costarricense: *Mitos y Realidades.*" *In Mitos y Realidades de la Democracia en Costa Rica,* eds. Yadira Calvo et al. San José, Costa Rica: DEI.

Rolbein, Seth. 1989. *Nobel Costa Rica: a Timely Report on Our Peaceful Pro-Yankee, Central American Neighbor.* New York: St. Martin's Press.

Romero, Rodia. 1991. "El Desarrollo Sostenible: Un Concepto Polémico." *Ciencias Ambientales* 8: 72–82.

Rosales, Johnny. 1997. "Reservas Naturales Privadas en Costa Rica y Un Estudio de Caso: Asociación Conservacionista de Monteverde y el Bosque Eterno de los Niños." Paper presented to 1st Latin American Congress on National Parks and Other Protected Areas. Santa Marta, Colombia. May 1997.

Rouse, Roger. 1991. "Mexican Migration and the Social Space of Postmodernism." *Diaspora* 1(1): 8–23.

Rovinski, Yvette. 1991. "Private Reserves, Parks, and Ecotourism in Costa Rica." In *Nature Tourism: Managing for the Environment,* ed. Tensie Whelan. Washington, DC: Island Press.

Rovira Mas, Jorge. 1989. *Costa Rica en los Años 80.* San José: Editoral Porvenir.

Rowell, Andrew. 1996. *Green Backlash: Global Subversion of the Environment Movement.* London: Routledge.

RSE (Reserva Santa Elena) 1990. "Perfil del Proyecto Ecoturistico-Investigativo Santa Elena de Monte Verde—Puntarenas." Project Proposal, Colegio Técnico-Profesional Agropecuario Santa Elena, Santa Elena, Monte Verde, Puntarenas, Costa Rica.

——— 1991. "What is the Santa Elena Reserve?" Placard in Interpretation Center, Reserva Santa Elena. Santa Elena, Puntarenas, Costa Rica.

n.d. *Santa Elena Reserve Self Guided Trail.* Santa Elena de Monte Verde, Costa Rica.

Rudel, Thomas and Bruce Horowitz. 1993. *Tropical Deforestation: Small Farmers and Land Clearing in the Ecuadorian Amazon.* New York: Columbia University Press.

Saez, Hector. 2003. "Pesticides in the 'Green Republic'" Econ-Atrocity series. Amherst, MA: U. Mass Amherst Center for Popular Economics.

Sachs, Wolfgang. 1996. "The Need for the Home Perspective." *Interculture* 2(1) (Winter 1996).

Salafsky, Nick. 2001. "Community-Based Approaches for Combining Conservation and Development." In *The Biodiversity Crisis: Losing What Counts,* ed. Michael J. Novacek. New York: The New Press.

Salazar, Roxana. 1993. *El Derecho a Un Ambiente Sano: Ecología y Desarrollo Sostenible.* San José, Costa Rica: Asociación Libro Libre.

Salazar, Roxana, et al. 1993. *Diversidad Biológica y Desarrollo Sostenible.* San José, Costa Rica: Fundación AMBIO.

Samper, Mario. 1985. "La Especialización Mercantil Campesina en el Noroeste del Valle Central. 1850–1900. Elementos Microanalíticos para un Modelo." *Revista de Historia,* Número Especial. Heredia, C.R. pp. 49–96.

Sanchez-Azofeifa, A. et al. 1997. "Building a remote sensing database and a geographic information system to monitor tropical deforestation and habitat fragmentation in Costa Rica.' Paper presented to the Association for Tropical Biology annual meetings, San José, C.R., June 1997.

Sandner, Gerhard. 1964. *La Colonización Agrícola de Costa Rica.* San José: Ministerio de Transportes, Instituto Geográfico de Costa Rica.

Scrimshaw, Nat, et al. 2000. "Conservation Easements in Monteverde: The Enlace Verde Project." In *Monteverde: Ecology and Conservation of a Tropical Cloud Forest,* eds. Nalini Nadkarni and Nat Wheelwright. Oxford: Oxford University Press.

Segura, Omar. 1992. *Desarrollo Sostenible y Políticas Económicas en América Latina.* San José, Costa Rica: DEI.

Seligson, Mitchell. 1980. *Peasants of Costa Rica and the Development of Agrarian Capitalism.* Madison: University of Wisconsin Press.

Selin, Helaine, ed. 2003. *Nature Across Cultures: Views of Nature and the Environment in Non-Western Cultures.* London: Kluwer Academic Publishers.

Sewastynowicz, James. 1986. "'Two-Step' Migration and Upward Mobility on the Frontier: The Safety Valve Effect in Pejibaye, Costa Rica." *Economic Development and Cultural Change* 1986: 731–53.

Shrader-Frechette, Kristin S. and Earl D. McCoy. 1993. *Method in Ecology: Strategies for Conservation.* Cambridge: Cambridge University Press.

—— 1995. "Natural Landscapes, Natural Communities, and Natural Ecosystems." *Forest and Conservation History* 39: 138–42.

Simberloff, Daniel and J. Cox. 1987. "Consequences and Costs of Conservation Corridors." *Conservation Biology* 1(1): 63–87.

Simmel, Georg. 1971 (1911). "The Adventurer." In *On Individuality and Social Forms: Selected Writings*, ed. D. Levine. Chicago: University of Chicago Press.

Solano, M. 1998. "Alerta sobre cambio en el clima." *La Nación* 4/2/98, p. 10.

Solera, A. and Tirso Maldonado, eds. 1988. *Desarrollo Socioeconómico y el Ambiente Natural de Costa Rica*. San José, Costa Rica: Fundación Neotrópica.

Solórzano, Raul and Jaime Echeverría. 1993. "Impacto Económico de la Reserva Biológica Bosque Nuboso de Monteverde." January 1993. Unpublished Internal Discussion Document. Centro Científico Tropical, San José, Costa Rica.

Soulé, Michael. 1985. "What is Conservation Biology?" *Bioscience* 35(11): 727–34.

Stepan, Nancy. 2001. *Picturing Tropical Nature*. Ithaca: Cornell University Press.

Stuckey, Joseph. 1988. *Kicking the Subsidized Habit: Viewing Rural Development as a Function of Local Savings Mobilization. The Santa Elena Case: Modification for the 1990s.* MBA Thesis, National University, San Diego, CA.

Sun, Marjorie. 1988. "Costa Rica's Campaign for Conservation." *Science* 239. 18 March 1988. pp. 1366–9.

Takacs, David. 1996. *The Idea of Biodiversity: Philosophies of Paradise*. Baltimore: Johns Hopkins University Press.

Tangley, Laurie. 1988. "Beyond National Parks." *Bioscience* 38(3): 146–7.

Taylor, John. 1994. *A Dream of England: Landscape, Photography and the Tourist's Imagination*. Manchester: Manchester University Press.

Terborgh, John. 1992. *Diversity and the Tropical Rain Forest*. New York: Scientific American Library.

Third World Network et al. 2001. "NGO Statement to Government Delegates at the UN." World Wide Web Site, accessed at: http://www.twnside.org.sg/title/eco4.htm on 9/21/01.

Thompson, Michael. 1999. "Security and Solidarity: An Anti-Reductionist Analysis of Environmental Policy." In *Living With Nature: Environmental Politics as Cultural Discourse,* eds. Frank Fischer and Maarten Hajer. Oxford: Oxford University Press.

Thrupp, Lori Ann. 1981. "El Punto de Vista Campesino sobre la Conservación." *Ceres,* Julio-Agosto 1991. pp. 31–4.

—— 1990. "Environmental Initiatives in Costa Rica: A Political Ecology Perspective." *Society and Natural Resources* 3: 243–56.

TIES (The International Ecotourism Society). 2004. "What is Ecotourism?" World Wide Web Site http://www.ecotourism.org/index2.php?what-is-ecotourism, accessed on 6/9/05.

TNC (The Nature Conservancy). n.d. "Conservation Abstract: Monteverde Cloud Forest." Arlington, VA.

Tobias, D. 1988. "Biological and Social Aspects of Eco-Tourism: The Monteverde Case. An Examination of Tourist Pressures in Monteverde, Costa Rica, with Recommendations for Strategies to Ameliorate Impacts." TRI Working Paper #34 Vol. 1. New Haven, Conn.: Yale School of Forestry and Environmental School.

Tobias, D. and R. Mendelsohn. 1991. "Valuing Ecotourism in a Tropical Rain-Forest Reserve." *Ambio* 20(2): 91–3.

Toledo, Victor. 2000. *La Paz en Chiapas: Ecología, Luchas Indígenas, y Modernidad Alternativa.* Mexico City: Ediciones Quinto Sol.

Tosi, Joseph. 1992. "Una Historia Breve de la Reserva Bosque Nuboso de Monteverde del Centro Científico Tropical – 1972–1992." Unpublished manuscript. San José, C.R.: Centro Científico Tropical.

Trejos, María Eugenia. 1990. "Nuevas Fórmulas de Consenso Social: El Ajuste Estructural en Costa Rica." In *Mitos y Realidades de la Democracia en Costa Rica,* eds. Yadira Calvo et al. San José, Costa Rica: DEI.

Tsing, Anna. 2000. "The Global Situation." *Cultural Anthropology* 15(3): 327–60.

Turner, Victor. 1974. *Dramas, Fields and Metaphors: Symbolic Action in Human Society.* Ithaca: Cornell University Press.

UNED/INBio, ed. 1994. *Del Bosque a la Sociedad/From Forest to Society.* San José, Costa Rica. Editorial UNED.

UNNGLS (UN Non-Governmental Liaison Service). 1997. *Implementing Agenda 21: NGO Experiences from Around the World.* Geneva: UNNGLS.

Urry, John. 1990. *The Tourist Gaze.* London: Sage Publications.

—— 1992. "*The Tourist Gaze* 'Revisited.'" *American Behavioral Scientist* 36(2): 172–86.

—— 1995. *Consuming Places.* London: Routledge.

Utting, Peter. 1993. *Trees, People and Power: Social Dimensions of Deforestation and Forest Protection in Central America.* London: Earthscan.

Vandermeer, John and Ivette Perfecto. 1995. *Breakfast of Biodiversity: The Truth About Rain Forest Destruction.* Oakland, CA.: Institute for Food and Development Policy.

Vargas, José Luis. 1995. "Principales Aspectos del Desarrollo de Monte Verde: 1920–1995." Presentation at II Foro Internacional Sobre Ecoalojamiento, Monte Verde, Costa Rica. October 1995.

Vargas, Mario. 1990. "Proyecto Historia Oral: Memorias Vivas de Monteverde." Unpublished interview transcript with Don Angel Villegas. Universidad Nacional Autónoma, Heredia, Costa Rica.

Viales, Ronny. 2002. "Ruralidad y Pobreza en Centroamérica en la Década de 1990. El Contexto de la Globalización y de las Políticas Agrarias 'Neoliberales.'" In *Culturas Populares y Políticas Públicas en México y*

*Centroamérica (Siglos XIX y XX)*, eds. F. Enríquez and I. Molina. Alajeula, Costa Rica: Museo Histórico Cultural Juan Santamaría.

Vieto, R. and J. Valverde. 1996. "Efectos de la Compra de Tierras con fines de Conservación." Unpublished manuscript. San José, Costa Rica: Recursos Naturales Tropicales, S.A.

Vivanco, Luis A. 2001. "Spectacular Quetzals, Ecotourism and Environmental Futures in Monte Verde, Costa Rica." *Ethnology* Vol. 40 no. 2, Spring 2001, pp. 79–92.

——— 2002a. "Seeing Green: Knowing and Saving the Environment on Film." *American Anthropologist* 104(4): 1195–1204.

——— 2002b. "Environmentalism, Democracy, and the Cultural Politics of Nature in Monte Verde, Costa Rica." In *Democracy and the Claims of Nature: Critical Perspectives for a New Century*, eds. Robert Pepperman Taylor and Ben Minteer. Lanham, MD: Rowman & Littlefield.

——— 2002c. "The International Year of Ecotourism in an Age of Uncertainty." Third World Network and Tourism Investigation and Monitoring Team, Clearinghouse for Reviewing Ecotourism Issue No. 23. World Wide Web Site, accessed at: http://www.twnside.org.sg.

——— 2003. "Conservation and Culture, Genuine and Spurious." In *Reconstructing Conservation: Finding Common Ground*, eds. Ben Minteer and Robert Manning. Washington, D.C.: Island Press.

Wallace, David Rains. 1992. *The Quetzal and the Macaw: The Story of Costa Rica's National Parks*. San Francisco: Sierra Club Books.

Wapner, Paul. 1996. *Environmental Activism and World Civic Politics*. Albany: SUNY Press.

Wearing, Stephen. 1992. "Ecotourism: The Santa Elena Rainforest Project." *The Environmentalist* 13(2): 125–35.

Wearing, Stephen and John Neil. 1999. *Ecotourism: Impacts, Potentials, and Possibilities*. Boston: Butterworth-Heinemann.

Weisgrau, Maxine. 1997. *Interpreting Development: Local Histories, Local Strategies*. Lanham, MD: University Press of America.

West, Paige and James Carrier. 2004. Ecotourism and Authenticity. *Current Anthropology* Vol. 45, no. 4 (Aug. 2004): 483–99.

Western, David and M. Pearl, eds. 1989. *Conservation for the Twenty-First Century*. Oxford: Oxford University Press.

Wheelwright, Nat. 2000. "Conservation Biology." In *Monteverde: Ecology and Conservation of a Tropical Cloud Forest*, eds. Nalini Nadkarni and Nat Wheelwright. Oxford: Oxford University Press.

Whelan, Tensie, ed. 1991. *Nature Tourism: Managing for the Environment*. Washington, DC: Island Press.

Williams, H. 1992. "Banking on the Future." *Nature Conservancy* May/June 1992, pp. 23–7.

Williams, Raymond. 1980. "Ideas of Nature." In *Problems of Materialism and Culture: Selected Essays.* London: Verso/NLB.

Wilson, Alexander. 1992. *The Culture of Nature: North American Landscape from Disney to the Exxon Valdez.* Cambridge, MA: Blackwell.

Wilson, Bruce. 1998. *Costa Rica: Politics, Economics and Democracy.* Boulder: Lynne Reiner.

Wilson, E.O., ed. 1988. *Biodiversity.* Washington, DC: National Academy Press.

Wolf, Eric. 1955. "Types of Latin American Peasantry: A Preliminary Discussion." *American Anthropologist* 57: 452–71.

Wood, Megan Epler. 2000. "The International Ecotourism Society: 10 Years on the CuttingEdge." *The Ecotourism Society Newsletter,* Second Quarter 2000. N. Bennington, Vermont.

WWF (World Wildlife Fund). 1986. Internal Memo from C. Freese to T. Lovejoy, dated March 12, 1986. File 6080. WWF—U.S., Washington, DC.

—— 1988. "World Wildlife Fund and the Conservation Foundation: A Quarter Century in Costa Rica, A Background Paper." WWF—U.S.: Washington, DC.

—— 1992. "Costa Rica: A WWF International Country Profile." February 1992. WWF—U.S.: Washington, DC.

YCI (Youth Challenge International) n.d. "Youth Challenge International." Flyer in Spanish, obtained in Santa Elena de Monteverde, Costa Rica.

Yearly, Steven. 1993. "Standing in for nature: the practicalities of environmental organizations' use of science." In *Environmentalism: The View from Anthropology,* ed. Kay Milton. ASA Monographs 32. London: Routledge.

Zuñiga V., A. 1990. "'Bosque Eterno de los Niños.'" *La Nación* 10/18/90. Viva section.

Zurick, David. 1995. *Errant Journeys: Adventure Travel in a Modern Age.* Austin: University of Texas Press.

# Index